Air Brake Technology

Third Edition

M N Homewood
I Eng MIRTE MIMI FIDiagE
Lecturer in Automobile Engineering at Woolwich College

Stanley Thornes (Publishers) Ltd

First published in 1977 by M N Homewood
Third edition published in 1991 by:
Stanley Thornes (Publishers) Ltd
Old Station Drive
Leckhampton
CHELTENHAM GL53 0DN
England

British Library Cataloguing in Publication Data
Homewood, Martin N.
Air brake technology.
I. Title
629.2

ISBN 0-7487-0564-3

Typeset by Tech-Set, Gateshead, Tyne & Wear.
Printed and bound in Great Britain at The Bath Press, Avon.

Contents

Acknowledgements

The author wishes to express his sincere thanks for the advice and assistance that has been rendered during the production of this book, by the following companies and organisations, and for their generosity in allowing the use of drawings and photographs of their products:

Bendix Ltd
BHVSI (Bendix Heavy Vehicle Systems Italia), formerly Magnetti Marelli
Robert Bosch Ltd
British Standards Institution (BSI)
V L Churchill Ltd
Clayton Dewandre Co Ltd
Paul Dahl Ltd
Department of Transport (Vehicle Inspectorate - Executive Agency)
ERF Ltd
Feeney & Johnson Ltd
Garphyttan Ltd
Grau Ltd, formerly Grau Girling Ltd
Hope Technical Developments Ltd
Iveco Ford Co Ltd
Knorr Bremze AG
Rubery Owen-Rockwell Ltd
VL Bus & Coach Co (UK) Ltd
WABCO Ltd
York Trailer Co Ltd.

Also published by Stanley Thornes (Publishers) Ltd:

J. Fairbrother	Principles and Practice of Vehicle Body Repair
V. A. W. Hillier	Auto Mate: How to Understand Your Car
	Fundamentals of Automotive Electronics
	Fundamentals of Motor Vehicle Technology
	Motor Vehicle Basic Principles
	Motor Vehicle Technology Associated Studies 1
	Motor Vehicle Technology Associated Studies 2
D. J. Leeming & R. Hartley	Heavy Vehicle Technology

Introduction

Industry, commerce and indeed the entire population of the UK depend upon heavy goods vehicles (HGVs) to deliver whatever they need. Daily, public service vehicles (PSVs) carry a large proportion of the population to and from work, and at other times play an important role in the tourist and leisure industries. They are seldom praised when they deliver goods and passengers promptly, and always criticised when they impede the progress of the car.

The very mention of an accident involving either a HGV or a PSV provokes a public outcry for yet more legislation. It is not always the fault of the drivers of such vehicles and equally it is not always the fault of the vehicle. However, there is still a minority of drivers and operators with lower standards than those expected, and they do a great deal of harm to the public image of heavy goods and public service vehicle operators.

The failure rate of HGVs, trailers and PSVs at their annual Department of Transport test is far too high. Consider just four common causes of test failure in 1989, which are as follows:

	HGV	*Trailer*	*PSV*
Service brake performance	8.0%	9.9%	5.4%
Secondary brake performance	7.2%	2.3%	3.9%
Parking system performance	5.9%	6.2%	4.0%
Braking system components	5.0%	3.7%	6.5%

Dealing with just one of these causes, we find that 585 816 HGVs underwent an annual test and that 8.0 per cent failed because the foot brake (service brake) did not meet the required standard. This means that over 46 865 vehicles were driven to the test station with the service brake in that condition. This example only relates to HGVs and indeed to just one aspect of the test. There may have been a few decisions which were borderline, but remember that 46 865 vehicles failed a test because of a defective service brake. An operator knows when a vehicle or trailer is going to be tested and indeed they know the requirements of that test. So, one cannot help wondering about the condition of such vehicles during the rest of the year.

It is not the aim of this publication to criticise, but merely to suggest that simple, regular, routine maintenance is the key to the successful operation of the good fleets that are using our roads today. Draining reservoirs, and checking lining thickness, push-rod travel, pressure build-up times and leakage, are all simple but very important tasks. Most people associated with workshops can quote actual expensive repairs and/or accidents resulting from the neglect of these checks by others.

The first edition of *Air Brake Technology* was published in 1977, the second edition in 1982, and these two editions were sold in 28 countries around the world. They were used by a great variety of people ranging from apprentices to fleet engineers, vehicle examiners to parts personnel, owner drivers to operators of large fleets, training schools and colleges. Indeed, copies are to be found on the premises of most air brake component and system manufacturers.

The aim of this publication is to provide an aid to those seeking to understand air brake components, systems, testing, maintenance and legal aspects, especially where they adopt the attitude that five minutes thinking time can save hours of stumbling around in the dark.

Air brakes

Hydraulic linkage has been common on light vehicles for many years and has the advantage of self-compensation, i.e. the vehicle should not pull to one side when braked, providing it is otherwise roadworthy. However, some form of assistance is necessary to reduce the driver's effort when applying such brakes.

Vacuum was widely used on heavy vehicles at one time, but there are limitations to such systems. Vacuum cannot exceed the pressure of atmosphere and indeed this pressure is seldom achieved. Therefore, the average vacuum assistance is equal to about 12 lbf/in^2 (83 kN/m^2, 0.8 bar), merely one-tenth of the air pressure used in the modern air brake system.

To achieve the same output force at the foundation brakes as a compressed air system, a vacuum system would need pistons or diaphragms in the actuators, with an area ten times greater than those in a comparable compressed air system.

Another disadvantage of a vacuum system is that with a simple leak, air enters the system and totally destroys that vacuum. Whereas with a compressed air system, although leaks are not desirable, a considerable leak can occur and yet the brakes will still be capable of stopping the vehicle.

For many years, HGVs and some PSVs have used compressed air as a linkage. Other PSVs have used air over hydraulics, i.e. a compressed air system that provides an actuator, with a push-rod to operate the master cylinder of an otherwise hydraulic linkage. Although manufacturers normally design braking systems to use compressed air at pressures of around 120 lbf/in^2 (827 kN/m^2, 8.3 bar), some systems have been produced where compressed air is stored in vehicle reservoirs at 250 lbf/in^2 (1723 kN/m^2, 17.2 bar). This high pressure, reduced before being supplied to brake chambers, is used to enable the number of storage reservoirs to be reduced, and so saves on cost and reduces the weight of the vehicle.

Vehicles often use compressed air to operate auxiliary equipment, such as the clutch, the gear box, a two-speed axle, the doors, the windscreen wipers and horns. Naturally, the air compressor must be capable of providing sufficient compressed air to exceed all likely needs of the braking systems and the auxiliary devices in the most adverse conditions.

Compressed air pressure must be controlled within the design limits of the system. The driver must be able to control the rapid application and rapid release of the brakes, and the brakes must be sensitive to the load being carried by the vehicle.

System and component design is biased to safety, incorporating warning devices, valves that prevent further air loss from a defective circuit, back-up circuits and brakes that operate automatically in the event of air loss. In spite of all these devices, poor maintenance or neglect can, and does, still cause disaster.

As a result of current legislation, no alterations should be made to any braking system before consulting the Vehicle Inspectorate of the Department of Transport or the technical department of any well-known brake manufacturer.

In the future, legislation is going to result in more and more vehicles being fitted with an electronic anti-lock braking system. Already some such systems have incorporated the additional feature of drive slip control, which enables the vehicle to be driven over very slippery surfaces without suffering wheel spin or losing traction.

Once such electronic systems become common-place, it is certain that other aspects of braking will also have electronic control. To take a simple example, a relay valve could be triggered electronically, giving an even faster signal from the control valve and thereby causing a faster response by the foundation brakes.

We can only wait and see how much additional electronic control is accepted by the vehicle manufacturers. There are arguments for and against the use of electronics on road vehicles. Nevertheless, it is worth remembering that large aircraft, both commercial and military, have been using electronic anti-lock braking systems very successfully for many years. Indeed, one aircraft has already been designed so that all mechanical aspects of that craft are fully controlled by electronic systems.

One only needs to look back at the very basic road vehicles of years ago, to appreciate how far design and performance have advanced so far, and to give us a hint of future possibilities.

M N Homewood

Codes for manufacturers' products

Throughout this publication the aim is to identify individual components to the reader. Often it helps if the manufacturer concerned is identified and, therefore, most components shown in this publication bear an identification code. The code (unique to this publication) uses letters to identify the relevant company, for example FJ represents Feeny & Johnson Ltd.

These codes are listed below. They are not in any order of preference, but in alphabetical order.

Manufacturers' codes

Bendix Heavy Vehicle Systems Italia	BI
Bendix Ltd	BW
Robert Bosch Ltd	BH
V L Churchill Ltd	CT
Clayton Dewandre Co Ltd	CD
Paul Dahl Ltd	PD
ERF Ltd	EF
Feeny & Johnson Ltd	FJ
Garphyttan Ltd	GN
Grau Ltd	GU
Haldex Ltd	HX
Hope Technical Developments Ltd	HE
Iveco Ford Ltd	ID
Knorr Bremze AG	KR
Rubery Owen-Rockwell Ltd	RR
VL Bus & Coach Co (UK) Ltd	VL
WABCO Ltd	WO
York Trailer Co Ltd	YK

Only one code is shown against most valves, although often an item is manufactured in a similar form by more than one company. However, there are many valves that are produced in a variety of forms, for example, left and right-hand gates, different ports, and with or without check-valve, and it would be both repetitious and boring to list all such differences. A limited number of valves shown, for example the load-sensing valve (MK1) and wick-type anti-freezer, are no longer fitted as original equipment to new vehicles, but there are plenty of such items still in use on older vehicles and naturally these older items are now more prone to failure or malfunction.

A full list of the postal addresses of the above companies is shown at the end of this publication.

1 Regulations that govern the design and use of air brakes on road vehicles

There are two main regulations that control the design and use of braking systems:

1 The Construction and Use Regulations; and
2 The European Economic Community Directives.

The British Construction and Use Regulations have been amended to include the requirements of the European Economic Community (EEC) and the Economic Commission for Europe (ECE) for new vehicles.

The requirements of the EEC and ECE, which have been incorporated into the Construction and Use Regulations, affect the design of air braking systems, whilst the use and maintenance of such systems are covered by the 'Use' section of the Construction and Use Regulations.

The Construction and Use Regulations

The Construction and Use Regulations are, as the name suggests, a part of the law that governs the design, manufacture, use and maintenance of the vast majority of motor vehicles that are used on the roads in this country.

The items covered by these regulations are numerous, ranging from the weight, length or width of a particular type of vehicle, to the maintenance of windscreen wipers. Naturally, braking systems have extensive coverage.

To interpret the regulations (although perhaps somewhat loosely) may be of assistance to the reader. However, it should be noted that the descriptions given in this chapter and throughout this publication should *not* be taken as legal definitions. A basic layman's explanation is given in most cases but where the text is differentiated this indicates that it is directly quoted material, mainly from EEC Directives.

All vehicles, used as goods vehicles, with an unladen weight over 1525 kg, are required to comply with the following braking requirements.

If first registered on or after 1 January 1968, the following efficiencies are necessary:

service brake (normally foot)	50 per cent;
secondary	25 per cent;
parking	16 per cent (i.e. capable of holding the vehicle on a gradient of 1:6.25).

If first registered prior to 1 January 1968, the following applies:

For two-axled vehicles –

service	45 per cent;
secondary	20 per cent;
parking	no requirement laid down.

For vehicles with more than two axles –

service	40 per cent;
secondary	15 per cent;
parking	no requirement laid down.

For trailers –

semi-trailer	40 per cent (of axle weight);
draw-bar trailer	50 per cent (of trailer weight).

The vehicle service brake is normally the foot brake. The secondary brake can be a hand-operated brake, or it can be that part of a service brake which remains operational in the event of a failure occurring in the service brake of a vehicle fitted with a split or dual-braking system.

Braking efficiency

The usual way of expressing the effectiveness of a braking system is to show it either as a part of gravitational pull (for example 0.5) or as 50 per cent (the figure showing the braking effort provided at the periphery of the road wheels compared with the weight of the vehicle and expressed as a percentage).

However the result of a test is expressed, some still believe that 100 per cent braking efficiency (BE) represents stopping 'dead'. This is *not* correct. A vehicle with 100 per cent BE would still require 30 ft (9.1 m) in which to stop, in addition to the reaction time that the driver takes to apply the brakes, when travelling at 30 m.p.h. (48 k.p.h.).

Although the Department of Transport test for private cars only requires the same 50 per cent as the foot brake for heavy goods vehicles (HGVs) or public service vehicles (PSVs), a much better figure should be obtained. The HGV and PSV are both liable to dislodge their load or passengers if excessively heavy braking takes place.

There are two ways of checking the efficiency of a braking system:

1 With a brake meter placed on the floor of the vehicle, the vehicle is driven at about 20 m.p.h. and then braked. These meters work on the pendulum principle, in that when the vehicle is braked, the pendulum inside the meter tends to continue moving forward. The amount that it travels is proportional to the degree of deceleration. The percentage of efficiency is read directly off the dial.
2 A 'rolling road' tester is used, as in Department of Transport test stations. The vehicle is checked on one axle at a time, with the rollers being rotated by an electric motor. The vehicle's brakes are applied, when requested by the machine operator, and the brake reaction at each wheel is read off gauges. Then the total braking effort for that vehicle is calculated by adding together all the readings. For example, if a two-axle vehicle has a gross weight of 5 tons (11 200 lbs) and if each wheel read 1400 lbs, this would total $1400 \times 4 = 5600$ lbs, or half the total weight of the vehicle, and thus the vehicle could be said to have a foot brake with an efficiency of 50 per cent.

   Using SI units, if a two-axle vehicle has a gross weight of 5 tonnes (5000 kg) and if each wheel read 625 kg, this would total $625 \times 4 = 2500$ kg, i.e. 50 per cent braking efficiency.

Whatever type of testing is carried out, the braking efficiency can be adversely affected by poor tyres or a slippery surface.

Important

Vehicles and trailers first manufactured on or after 1 October 1982 and first used on or after 1 April 1983 must be fitted with a braking system which complies with the EEC directives.

The ECE Regulation 13 has now been included in UK Regulations and vehicles can be 'type approved' to this standard, if desired, as an alternative to the EEC Standard.

The following vehicles are exempt from this requirement:

- agricultural trailers;
- locomotives;
- motor tractors;
- vehicles which are incapable, by reason of their construction, of exceeding a speed of 25 k.p.h., on the level, under their own power;
- works trailers;
- works trucks;
- public works vehicles (Post Office, Highway Authority, etc);
- trailers drawn exclusively by any vehicle listed above.

The British Leyland Atlantean Bus, if manufactured on or after 1 October 1983 and first used on or after 1 October 1984, is also exempt from complying with these regulations. Likewise certain other double-decked buses, first used before 1 October 1983, are not affected by these Directives.

There are many unusual vehicles, for example articulated buses or vintage lorries. Thus, it is essential to check the requirements for such vehicles. Information relating to 'type approval' of vehicles or components may be obtained from:

The Vehicle Certification Agency,
1 The Eastgate Office Centre,
Eastgate Road,
Bristol, BS5 6XX
Telephone: 0272 524125.

EEC Directives

The Economic Commission for Europe (ECE), an agency of the United Nations, issues regulations, which may voluntarily be used by countries within the ECE. The EEC Directive on braking has been formulated around the ECE regulation on braking.

As with all acts and regulations, new Directives are constantly being produced. Directive 71/320/EEC is the bulkiest so far and is updated by later Directives. These include 74/132/EEC, 75/524/EEC, 76/770/EEC, 79/489/EEC, 85/647/EEC and 88/194/EEC.

No attempt has been made to give a legal summary of the Directives, but rather an insight for the layperson as to the way that the EEC regulations affect motor vehicles in the UK is provided.

The following may assist when reading either the following summary or the actual Directives:

Common usage	*EEC Directives*
Weight	Mass
Maximum weight	Maximum mass
Force	Units = newton (N)
1 kg or 1 kgf	10 N
Torque	Units = newton metres (Nm)
1 mkg or 1 mkgf	10 Nm
Pressure	Units = bar
1 kg/cm^2 or 1 kgf/cm^2	1 bar

Referring to 71/320/EEC (as amended)

Article 1: This lists the categories of vehicles. It defines 'vehicle' as any motor vehicle falling within one of the international categories listed, with or without bodywork, with a minimum of four wheels, having a speed capability of over 25 k.p.h. (approximately 16 m.p.h.) and includes trailers. (Exemptions are agricultural tractors, machinery, public works vehicles and those vehicles that run on rails.)

Category M Passenger vehicles with at least four wheels or three wheels when the maximum weight is over 1 tonne.

M1 Passenger vehicles with up to eight seats, plus driver's seat.

M2 Passenger vehicles with a greater number of seats than in M1 but not exceeding 5 tonnes maximum weight.

M3 Passenger vehicles with a greater number of seats than M1 but where the maximum weight exceeds 5 tonnes.

Category N Goods vehicles with at least four wheels or three wheels with a maximum weight over 1 tonne.

N1 Goods vehicles having a maximum weight not exceeding 3.5 tonnes.

N2 As N1, but with a maximum weight of over 3.5 and not exceeding 12 tonnes.

N3 As N1, but exceeding 12 tonnes maximum weight.

Category O Trailers (including semi-trailers).

O1 Trailers with a maximum weight not exceeding 0.75 tonnes.

O2 As O1, but with a maximum weight of over 0.75 and not exceeding 3.5 tonnes.

O3 As O1, but with a maximum weight of over 3.5 and not exceeding 10 tonnes.

O4 As O1, but with a maximum weight of over 10 tonnes.

Article 2: This lays down that no member state can refuse to grant either EEC or National Type Approval of a vehicle on grounds relating to the braking equipment, if the devices comply with Annexes 1 to 8 (the regulations governing the 'construction' aspect of brake equipment).

This instruction is most useful to the vehicle manufacturer in the UK, as once the vehicle design has achieved EEC Type Approval, it cannot be refused (on design grounds) by any member state of the EEC.

Annex I:

1.5 'Transmission' is a word defined in these Directives as meaning:

The combination of components situated between the control and the brake and connecting the two operationally. The transmission may be mechanical, hydraulic, pneumatic, electrical or mixed.

Without the above explanation, there may well be confusion, especially with items 2.2.12.2 and 2.2.12.3.

1.18 'Inter-urban motor coach' means a vehicle designed and equipped for inter-urban transport, having no spaces specifically intended for standing passengers, but able to carry, for short distances, passengers standing in the gangway.

1.19 'Long-distance touring motor-coach' means a vehicle designed and equipped for long-distance journeys, arranged to ensure the comfort of its seated passengers and which does not carry standing passengers.

2.1.3 Pneumatic connections between motor vehicles and trailers.

2.1.3.1 In the case of a braking device operated by compressed air, the pneumatic link with the trailer must be of a type with two or more lines. However, in all cases, all the requirements of this Directive must be satisfied by the use of only two lines. Shut-off devices which are not automatically actuated shall not be permitted. In the case of articulated combinations, the flexible hoses shall be part of the drawing vehicle. In all other cases, the flexible hoses shall be part of the trailer.

2.2.1.2.1 There must be at least two controls, independent of each other and readily accessible to the driver from his normal driving position. For all categories of vehicles, except M2 and M3, every brake control (excluding a retarder control) shall be designed such that *it returns to the fully-off position when released*. This requirement shall not apply to a parking brake control (or that part of a combined control) when it is mechanically locked in an applied position.

2.2.1.2.4 If the service and secondary brakes share the same control, the parking brake must be so designed that it can be operated when the vehicle is in motion.

2.2.1.4 If there is a failure in a part of the transmission of the service brake, a sufficient number of wheels must still be braked by the action of the control of the service brake, so that the residual efficiency of the service braking equipment is equal to 30 per cent of the efficiency for the vehicle (loaded).

For unladen vehicles the percentage is 25 per cent for vehicles in groups M1, M2, N1 and N2, and 30 per cent for groups M3 and N3.

To meet this requirement, a dual system can be used, split under the foot, provided that one half can meet the residual performance, as shown above. If each half of the system has been designed to comply with the above requirements, it can also be used for the secondary brake, providing that each half of the system can meet the performance required for the secondary on its own. Alternatively, the secondary system can be controlled by a hand valve.

2.2.1.5 If one section of the complete system fails, the feed to the rest of the system must continue to be operable, as necessary, to stop the vehicle with the efficiency required for the secondary brake.

To meet this requirement the UK manufacturers have introduced into their layouts a valve described variously as a single-circuit charging valve or a protection valve (depending upon the maker), or if two or more such valves are contained in the same casting, they are known as a multi-circuit charging valve or a multi-protection valve.

2.2.1.11 Wear on the brakes must be easily compensated by means of a system of manual or automatic adjustment. In addition, the control and the components of the transmission and of the brakes must possess a reserve of travel such that, when the brakes become heated or when the brake linings have reached a certain degree of wear, effective braking is ensured without an immediate adjustment being necessary.

2.2.1.12.1 The filling ports of the fluid reservoirs must be readily accessible; in addition, the containers of reserve fluid must be so made that the level of the reserve fluid can be easily checked without the containers having to be opened. Where this last condition is not fulfilled, a warning light must indicate to the driver when the reserve fluid falls to a level liable to cause a failure of the braking device. The driver must be able to check easily whether the light is functioning properly.

2.2.1.12.2 The failure of a part of a hydraulic transmission shall be signalled to the driver by a device comprising a red tell-tale lamp lighting up not later than on actuation of the control. However, a device comprising a red tell-tale lamp lighting up when the level of the fluid in its reservoirs falls below the value specified by the manufacturer, is admissable. The tell-tale lamp shall be visible even by daylight; the satisfactory condition of the lamp must be easily verifiable by the driver. The failure of a component of the device shall not entail total loss of effectiveness of the braking device in question. (See item 1.5 above, regarding the interpretation of 'hydraulic transmission'.)

2.2.1.13 A low-pressure, optical or acoustic, warning device, shall be fitted, to give a signal when pressure in any part of the system falls to a value at which it is still possible to apply the service brake control for four full actuations, and then a further (fifth) application to the value of secondary braking performance.

2.2.1.19 In the case of a vehicle authorised to draw a category O3 or O4 trailer, the braking devices must satisfy the following conditions:

2.2.1.19.1 When the secondary braking device of the drawing vehicle is actuated, there must also be a graduated braking action on the trailer.

2.2.1.19.2 In the event of failure of half of a dual system, the unaffected part(s) must be able to actuate the trailer brakes, in a graduated manner. (Manufacturers have incorporated multi-relay valves into their circuits for this purpose.)

2.2.1.19.3 In the case of a break or leak in one of the air supply pipes to the trailer, which is not severe enough to apply the trailer brake, via emergency line and relay emergency valve, it must be possible for the driver to apply, or partially apply the trailer brakes by means of either the service, secondary or separate control.

2.2.1.22 Certain motor vehicles shall be fitted with anti-lock devices in accordance with Annex X, as shown in the following table:

Vehicle		*Maximum mass*	*Anti-lock category*
Category	*Description*		
M3	Inter-urban motor coaches and long-distance touring motor coaches	12 tonnes	1
N3	Motor vehicles authorised to tow category O4 trailers	16 tonnes	1

Note: See Items 1.18 and 1.19 for the definition of the above coaches.

2.2.1.23 If motor vehicles not mentioned in item 2.2.1.22 are fitted with anti-lock devices, they shall comply with the requirements of Annex X.

2.2.2.13 Trailers of category O4 shall be fitted with anti-lock devices in accordance with Annex X.

2.2.2.14 If trailers not mentioned in item 2.2.2.13 above are fitted with anti-lock devices, they shall comply with the requirements of Annex X.

Note: Annex X deals with requirements applicable to tests, by manufacturers, for vehicles equipped with anti-lock devices.

Annex III:

4.1 and 4.2 On motor vehicle and trailers each independent braking system requires a pressure test connection (complying with ISO Standard 3583:1984), to be fitted as close as possible to the brake cylinder which is least favoured as far as reaction time is concerned.

Annex IV:

2.5.1 Charging reservoirs from zero to normal working pressure should be achieved, in the case of a rigid vehicle in 8 minutes, and in the case of a combination in 11 minutes.

3.1 and 3.2 Motor vehicles and trailers require a pressure test connection (complying with ISO Standard 3583:1984), to be fitted as close as possible to the least favourable reservoir fitted to that vehicle or trailer.

Annex V:

2.1 A spring brake should not be used as a service brake.

2.3 The spring compression chamber (spring brake) supply circuit must include an energy reserve which does not supply any other equipment. This does not apply when the springs can be kept compressed by using two or more independent systems. This item does not apply to trailers.

2.4 Spring brake systems on vehicles must be designed so that it is possible to apply and release the brakes at least three times, if the initial pressure in the spring compression chamber is equal to maximum design pressure.

With trailers it must be possible to release the spring brakes at least three times, after the trailer has been uncoupled. The pressure in the supply line being at 6.5 bar before uncoupling.

2.7 With motor vehicles fitted with spring brakes, drawing trailers fitted with air brakes, when the spring brakes operate automatically, the trailer brake must also be caused to operate.

Note: There are several ways of achieving this by the use of relay valves in the circuits.

3.1 Spring brakes must be designed so that in the event of a failure it is possible to release them without using their normal control. This device can be mechanical, pneumatic, etc.

3.2 If the device referred to in 3.1 above is a spanner or tool, it must be kept on the vehicle.

Annex IX

3.1.2 A towing vehicle to draw trailers of Category O3 and O4 must have pressures between 6.5 and 8 bar at the coupling head of the supply line and between 6 and 7.5 bar at the coupling head of the control line.

7.1 Vehicles, other than those of category M1, with mechanically-controlled load-sensing require a 'plate' to be affixed, showing the useful 'travel' of that device, between the laden and unladen positions. Any further information necessary for this setting must also be shown.

8 Vehicles and trailers fitted with a load-sensing device, which modulates the air pressure that is to be delivered to the brake chambers, must be fitted with pressure test connections in the pressure line, upstream and downstream of the device. The test connections must conform with ISO Standard 3583:1984.

Annex X:

The purpose of this Annex is to define the required performances for braking systems with anti-lock devices fitted to road vehicles.

Perhaps the most reassuring item is 4.5 below, which should do much to allay fears arising from claims that emergency radios and similar devices affect the performance of anti-lock systems.

4.5 The operation of the device must not be adversely affected by magnetic or electrical fields.

2 Metrication

As a result of a recommendation made by the International Standards Organisation (ISO) in 1960, that to assist international trade a single set of measurements was necessary, the UK has been co-operating with this aim. The measurements now in use in the UK are contained in the *Systéme International D'Unités*, which is abbreviated internationally to the SI System.

Manufacturing industry has changed from imperial units to SI units, for example from feet and inches to metres and millimetres, although some maintenance workshops still use a mixture of imperial and SI units. Several alternative SI units are still used by different branches of industry, including bar, pascal and kilonewton/metre^2. Undoubtedly, all of these units have provoked comment, but within a few years they will become as commonplace and accepted as pounds per square inch (lb/in^2). Can we really recognise what 100 lb/in^2 means or are we just accepting a familiar unit and noting that the figure is high? However, like it or not, the SI System is here, and since 1977 the City & Guilds Part Three Examination 'Compressed Air Brakes' has used SI units. Consequently, it is important that some conversion of units is explained.

Other units in common use are:

metric tonne	= 1000 kilogram; 2205 lb
ton	= 2240 lb
bar	= 10^5 N/m^2; 100 kN/m^2 (or approximately 14.5 lbf/in^2, i.e. slightly less than a standard atmosphere)
pascal	= Pa (an alternative and equivalent to N/m^2).

The newton is a unit of force named after Sir Isaac Newton, and the symbol used is N. It is the force which when applied to a mass of one kilogram gives an acceleration of one metre per second, every second.

$$\text{Thus 1 newton} = 1\ \text{kg} \times 1\ \text{m/s}^2$$

The SI System is aided by the kilo- (10^3, i.e. $\times 1000$) and milli- (10^{-3}, i.e. $\div 1000$). However, care should be taken to use the correct small or capital letters in abbreviations, for example N = newton, m = metre, M = mega- (one million).

As the pascal and newton/metre^2 are such small units, they have little practical use (approximately 7000 Pa = 1 lbf/in^2), thus industry has the choice of kPa or kN/m^2 (approximately 7 kPa or 7 kN/m^2 = 1 lbf/in^2), and bar (approximately 14.5 lbf/in^2).

Imperial unit	*SI unit*
One inch (1 in)	2.54 cm (centimetre); 25.4 mm (millimetre)
One foot (1 ft)	0.3048 m (metre); 304.8 mm
1 inch^2 (1 in^2)	6.451 cm^2; 645.16 mm^2
1 inch^3 (1 in^3)	16.387 cm^3
1 lbf	4.448 22 N (newton)
1 lbf/in^2	6.894 76 kilonewton/metre^2 (kN/m^2); 6894.76 N/m^2
1 mile	1.609 344 km (kilometre)
5 miles	8 km

3 The language of pneumatics

In all specialised fields of industry, expressions, American terms and even jargon creep in, until specifications and descriptions sound like a foreign language. The pneumatics industry is no exception. Throughout this publication the aim is to keep to everyday language, as far as possible. However, where this is not possible and to help in general understanding, a short list of common terms is as follows.

ABS The abbreviation for the German term anti-blockier system, i.e. anti-lock or anti-skid braking system.

Air-limiter valve A name used by some vehicle manufacturers for a governor valve.

Air over hydraulics This refers to hydraulic braking systems, where the driver's control (foot valve treadle) routes compressed air, via that valve, to an actuator (or servo-device), which in turn causes its mechanical push-rod to activate the hydraulic master cylinder. From this point, the system is identical with an orthodox hydraulic braking system.

Anti-compound To prevent *both* air pressure and spring pressure being applied at the same time.

Anti-compound valve Such valves are fitted to prevent the combined forces of spring brakes and conventional air brakes acting simultaneously, thereby causing damage to push-rods, slack adjusters and foundation brakes. Differential-protection valves and double-check valves come within this description.

Anti-lock Electronic sensing is used to prevent road wheels from locking when the vehicle is being braked. This is achieved by releasing the affected brake just as that wheel and tyre is about to lock or skid, and then very quickly reapplying that brake. This is repeated as often as necessary, giving a very rapid pulsing effect of the brake(s) being applied, and thereby achieving the shortest possible stopping distance for that vehicle and the particular road surface.

ASR The abbreviation for the German term antriebs schlupf reglung, i.e. traction control system. The ability to drive/pull away on icy surfaces.

Automatic drain valve A device fitted to reservoirs, wet tanks and condensers, to enable moisture in the air within to be ejected automatically, and to remove the need for manual draining of that unit.

Auxiliary line *See* Secondary line.

Balanced condition *See* Lapped condition.

BDC The abbreviation for bottom dead centre, i.e. the lowest point that a component can reach.

Brake actuator A unit where the compressed air moves a mechanical component, which in turn applies the brakes, via mechanical linkage.

Brake chamber This is another name for a brake actuator.

Check valve Also known as a *non-return valve*, this unit allows air pressure to travel in one direction only.

Choke This is a small drilling within a valve to restrict the air flow. It is used particularly to provide a tell-tale pressure reading within a control valve, necessary to achieve a lapped or balanced condition.

Circuit-protection valve This type of valve, produced as a 'single' or 'multi' is the improved version of the check valve or non-return valve. It was introduced to comply with an EEC Directive which states that if one part of a braking system should fail, the unaffected part of that system must be capable of stopping the vehicle with at least the efficiency that is required for the secondary system. These values are produced in two forms, one with a feedback capability (so that the pressures in certain reservoirs can be equalised, whilst the system is being charged) and another type that does not have the feedback capability. Another name used for this range of valves is *system-protection valves*.

Compounding of forces This term describes the result of two independent forces acting upon the same unit simultaneously, for example the spring and the air pressure in the conventional spring-brake chamber. The result would be the overloading of mechanical parts, with resultant damage. To overcome this, anti-compounding valves are fitted.

Control line An alternative name for the service (yellow) flexible hose between drawing vehicle and trailer.

Couplings: (C and CA-types) These are types of quick-release couplings (usually fitted with a self-sealing valve that acts when they are separated) used to connect the flexible hoses between the drawing vehicle and its trailer. The CA-type is larger than the C-type but providing orthodox components are used, tractor or trailer lines cannot be wrongly connected.

Couplings (palm type) A type of coupling used for connecting flexible brake hoses between the drawing vehicle and its trailer. When the two hoses are connected, the device resembles clasped hands and is sometimes called *glad hand*. Some versions are produced with self-sealing valves.

Differential-protection valve One type of anti-compounding valve, fitted into spring-brake circuits. Also known as an *overload-protection valve*.

Double-check valve This valve allows pressure, from two sources, to supply a brake actuator, but only allows one source access to that valve at a time. This unit can be used in spring brake circuits to prevent compounding.

Downstream Meaning beyond a particular point in the system, for example the brake chambers are downstream from the foot valve.

Emergency line This line feeds the trailer reservoir, through a relay emergency valve. If this line breaks or fails continually to contain air pressure, the relay emergency valve will automatically cause the trailer brakes to be applied, by supplying compressed air from the trailer reservoir. The usual colour for this hose is red, although this is not a legal requirement. Also known as a *supply line*.

Foot valve This is the valve attached, either directly or indirectly, to the driver's footbrake treadle.

Foundation brakes The conventional brakes, i.e. shoes and lining, drum, camshaft, etc., that the compressed air system activates.

Glad hand *See* Couplings (palm type).

Governor This valves reacts to the pressure within the reservoir, so that, at a predetermined pressure, a signal pressure is delivered to the compressor, which causes it to run light and not deliver further pressure to the reservoir until needed, i.e. when the reservoir pressure drops.

Graduable Meaning the same as progressive. The important point is to distinguish between an on/off valve and one that is under the control of an operator, such as a foot or hand valve, which can be controlled, i.e. is graduable.

Hand valve A valve to which the driver's hand control lever is attached, which gives graduable control, usually over the secondary and/or parking brakes.

Inshot In a braking system fitted with a load-sensing valve which has an inshot facility, a light application of the foot valve treadle causes a shot of air to be delivered to the brake chambers. This is to overcome the initial stiffness of the foundation brakes and also to prevent glazing of the brake lining material.

Inverse air As the term 'inverse' implies, this means upside-down, or in this case the reducing of air pressure. This particularly applies to spring-brake systems, as air is used during 'normal' driving to hold the spring pressure off. By the reduction of the air pressure in the spring chamber, the spring is allowed to extend to its normal length, applying the brakes. Control over the inverse air is normally achieved by the use of a special hand valve.

Lapped condition Control valves, such as the foot or hand valve, would make a vehicle difficult to handle properly if they were solely on/off switches. This would be the situation should maximum reservoir pressure be allowed to pass to the brake chambers, irrespective of whether a light or heavy application is intended. To overcome this problem, a small quantity of incoming air is diverted to act against the graduating (progressive) spring beneath the foottreadle or hand lever. As the air pressure builds up within the valve, it matches and overcomes the opposing mechanical force of the spring. Consequently, the spring will move up, allowing the inlet valve to close and yet leaving the exhaust valve closed, but with the brakes still applied. This is also known as the *balanced condition*.

Light-laden valve Some manufacturers use this name for the *load-sensing valve*.

Load-sensing valve This valve is usually fastened to the chassis, but with a link connected to the axle so that the amount of load carried by the vehicle registers in the valve according to the distance between axle and chassis. The object is to apportion the amount of braking according to load, i.e. maximum on a fully-laden vehicle, but limited braking on a lightly-laden vehicle. (The aim is to prevent skidding and jack-knifing.)

Lock actuator This is a brake actuator that can be used for both service and parking brakes and, in most cases, for secondary braking. During normal driving the lock rollers are held away from the push-rod by air pressure. If the air pressure in this lock line is exhausted, the brakes are prevented from releasing by the rollers. This system should not be confused with spring brakes.

Non-return valve *See* Check valve.

Overload-protection valve Another name for a *differential-protection valve*.

Overflow valve *See* Pressure-protection valve.

Palm coupling *See* Couplings (palm type).

Predominance This gives a lead to the trailer brakes to achieve greater balance with the brakes of the drawing unit. When the foot valve is applied and a given pressure is delivered, predominance occurs when this pressure is exceeded at the coupling head.

Pressure-limiting valve *See* Pressure-reducing valve.

Pressure-protection valve As the name suggests, this valve is installed in a pipeline to ensure that air does not pass beyond the valve point, unless the preceding section has achieved the designed pressure. These valves are installed between reservoirs, to ensure that the service reservoir has priority, or in the emergency line to the trailer, so that the tractor has priority, should the line beyond this valve lose pressure. Also known as an *overflow valve*.

Note: One type of pressure-protection valve has a feedback capability, i.e. although it prevents the flow of compressed air beyond the valve, it will allow the flow back, from beyond, of air that is at a higher pressure. This return air can be used in the protected circuit.

Pressure-reducing valve This valve is fitted into a system to reduce the pressure beyond the valve point to suit the components fitted, for example air-operated windscreen wipers, which do not require the full system pressure. Also known as a *pressure-limiting valve*.

Pressure regulator or pressure-regulator valve Basically, this has the same role as a pressure-protection valve, i.e. it safeguards the system prior to the valve, until the design pressure is reached, at which stage the valve opens and allows pressure to pass through to other sections of the system. (Confusion may occur due to manufacturers referring to an unloader valve as a pressure regulator.)

Pressure-relief valve *See* Safety valve.

Quick-release valve To allow the air released from a brake chamber to exhaust to atmosphere, close to that chamber.

Regulating valve To vary the amount of braking force applied to the brakes of different axles.

Relay valve To speed up the application and release of the brakes, by bringing a supply of air pressure close to the brake, which is triggered by a small volume (signal) of air from the control valve. An exhaust is also fitted to this valve.

Relay-emergency valve This is a relay valve with an additional feature. It will cause the trailer brakes to apply automatically, should the supply in the emergency line fail.

Safety valve A safety valve is incorporated into all air brake systems, to prevent automatically the build up of excessively high pressures that would otherwise occur in the event of the unloader or governor valve becoming defective. Also known as a *pressure-relief valve*.

Secondary line This is the secondary or auxiliary circuit of the vehicle, usually hand-operated. This can be used in the event of the service brake failing. The usual colour of the secondary flexible hose connection to the trailer is blue, although this is not a legal requirement. This is also known as the *auxiliary line*.

Service line The main braking circuit of the vehicle, which on orthodox vehicles is the footbrake system. It can also be a description of the flexible hose connecting the drawing vehicle and trailer. This is usually yellow, although this is not a legal requirement.

Slack adjuster These are fitted to the cam shaft of the foundation brakes to provide an external adjuster to enable easy adjustment, and thus to compensate for brake lining wear.

Spring-brake chamber For normal driving, compressed air holds a very powerful spring within the chamber in a compressed state. For parking, or in the event of that circuit failing, the air holding the spring is exhausted and the brakes are held 'on' by the power of the spring.

Supply line An alternative name for the *emergency line*.

Susie hose (trade name used by Dana Ltd) The pre-coiled plastic hoses used to connect the braking systems of the drawing vehicle and the trailer.

System-protection valve An alternative name for a *circuit-protection valve*.

TDC The abbreviation for top dead centre, i.e. the highest point that a component can reach.

Three-line system Where there are three flexible connections (service, secondary and emergency) to link the braking systems of the drawing vehicle and the trailer.

Treadle The footbrake pedal of the conventional air brake system is a platform known as a treadle.

Two-line system With the EEC-type system, this entails having a service and an emergency line, although it is still possible to use the secondary control on the drawing vehicle and obtain, in addition to secondary braking on the tractor, braking on the trailer via the service line. However, only single-diaphragm brake chambers can be used, unlike the three-line system where either double or triple-diaphragm chambers can be used.

Unloader valve This valve controls the pressure of the air in the reservoir by exhausting the supply coming from the compressor, before it reaches the reservoir.

Upright air This is normal compressed air, for example where a foot or hand valve system uses compressed air to apply the brakes. Conversely, if brakes are applied by reducing air pressure, i.e. spring brakes, this is known as *inverse air pressure*.

Upstream Any point prior to a particular point in the system, for example the foot valve is upstream from the brake chamber.

Variable-load valve The purpose is the same as the load-sensing valve, although the internal mechanism is different.

4 An explanation of the relationship between pressure, area and force

Most of the components of an air braking system are operated by valves, springs, diaphragms, pistons and by the variation of pressure. It is essential that the fundamental principles should be explained, in basic terms, so that more complex components can be readily understood.

Simple formulae

The mention of a formula will tempt many readers to turn over the page, but all we have to consider is *Force* = *Pressure* × *Area*, or just using the letters, $F = P \times A$.

The size (area) of the diaphragm inside a brake actuator depends upon the part number and the type of actuator. A type 30 has a working area of 30 square inches (30 in^2) whereas a type 24 has a working area of 24 square inches (24 in^2). Naturally, if 100 lbf/in^2 pressure is applied, the type 30 will provide a greater output force on the push-rod than the type 24.

Pressure

Pressure is the amount of weight that acts upon each unit of area. In braking systems, until the advent of SI units, pressure was measured in pounds per square inch, psi or lbf/in^2. A common figure of 100 lbf/in^2 is equivalent to 100 pounds acting on each square inch of the item or component that is being described.

Now, the SI unit, the **newton per metre²** is used, but the kilonewton per metre² is more practical. 1 lbf/in^2 = 6.894 kN/m^2; therefore 100 lbf/in^2 = 689 kN/m^2 approximately.

Although it can be easier to use square millimetres (mm^2) for the area, it can cause mistakes in calculations when converting to square metres (m^2), as 1 m^2 = 1 000 000 mm^2.

Atmospheric pressure

This is the weight of air that bears down, or the pressure that air exerts, upon the surface of the earth. At sea level this is 14.7 lbf/in^2. In SI units, the **bar** is roughly the same (14.5 lbf/in^2).

Gauge pressure

Normal pressure gauges only show the pressure *above* atmospheric pressure.

Explanation of diagrams

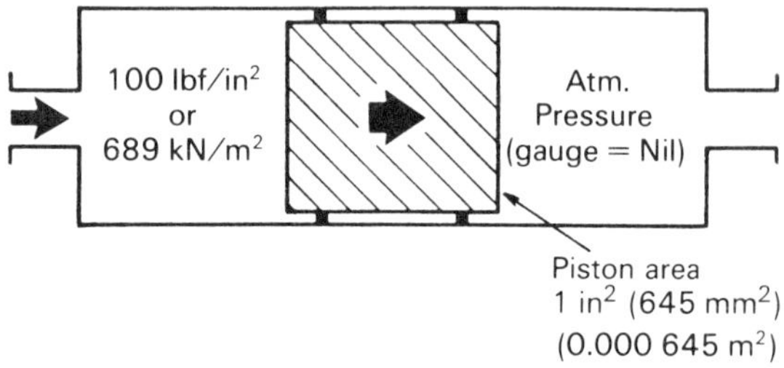

Figure 4.1 Force = Pressure X Area

Figure 4.1: A piston is located in a smooth bore and a pressure of 100 lbf/in^2 is fed into the left-hand end of the barrel. Obviously as there is only atmospheric pressure at the right-hand end, the piston will move towards the right. As force equals pressure × area, the force (ignoring friction) will be 100 × 1 = 100 lbs.

Similarly, the SI units will be 689 kN/m^2 × 0.000 645 m^2 = 0.444 405 kN. However, by reverting to newtons instead of kilonewtons, the figure becomes more practical, i.e. 0.444 405 × 1000 = 444 newtons (approximately).

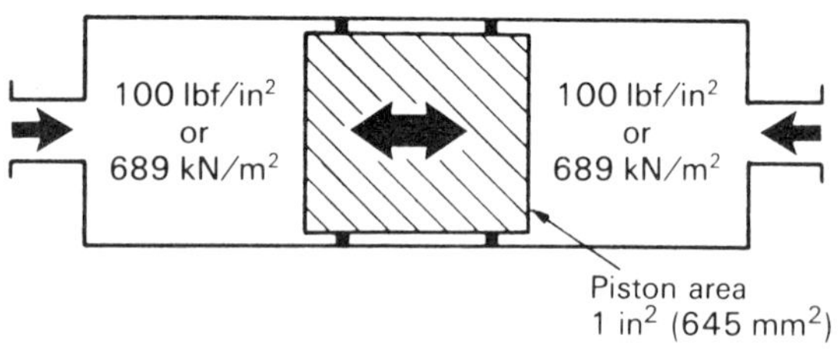

Figure 4.2 Equal pressure and equal area

Figure 4.2: As the pressure is equal from both directions, the piston will not move.

Figure 4.3: Although the area is the same, the pressure from the right is far less than that entering from the left. Using *Force* = *Pressure* × *Area*:

Left = 100 × 5
Right = 10 × 5.

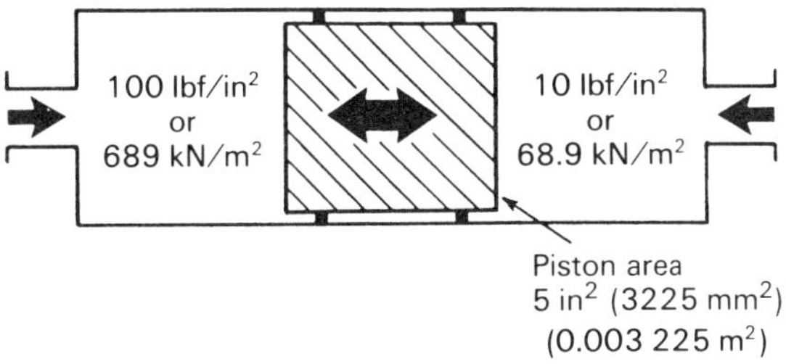

Figure 4.3 Unequal pressure and equal area

Therefore, the piston will move to the right with a force equal to $500 - 50 = 450$ lbs.

Using SI units:

$$\text{Left} = 689 \times 0.003\,225 = 2.222\,025 \text{ kN}$$
$$= 2222 \text{ N}$$
$$\text{Right} = 68.9 \times 0.003\,225 = 0.222\,202\,5 \text{ kN}$$
$$= 222 \text{ N}$$

Therefore, the piston will move to the right with a force equal to $2222 - 222 = 2000$ N.

Figure 4.4: This diagram shows a piston that has a large face area at the right-hand end and a small face area at the left-hand end. Referring to the formula for force (*Force* = *Pressure* × *Area*), it will be seen that, with an equal pressure applied to each end, the piston will move towards the left:

Right-hand
$= 100 \times 10 = 1000$ lbf
Left-hand
$= 100 \times 5 = 500$ lbf

500 lbf, the movement to the left.

Using SI units:

Right-hand
$= 689 \times 0.006\,450 = 4.444\,050$ kN
Left-hand
$= 689 \times 0.003\,225 = 2.222\,025$ kN

2.222 025 kN or 2222 N.

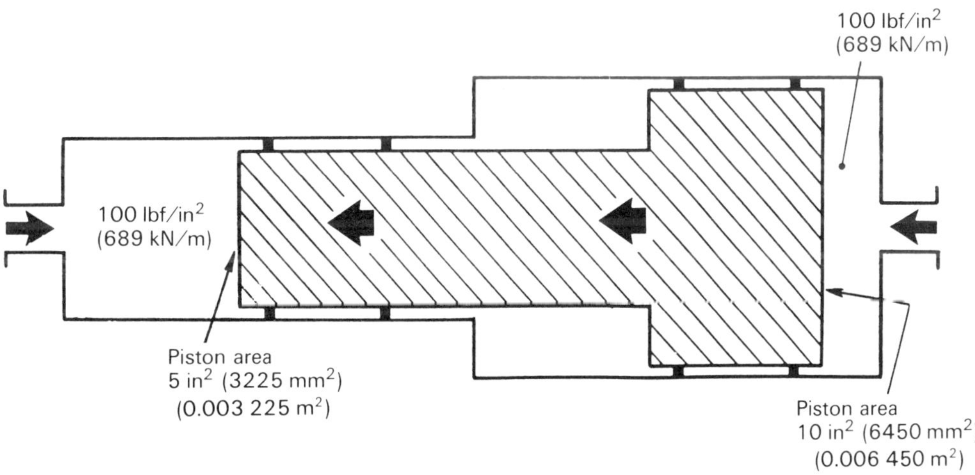

Figure 4.4 Equal pressure and unequal area

5 The principles of progressive valves

The operation of progressive valves (graduable action)

As shown in Figure 5.1, the base of the unit is connected to a supply of compressed air, but sealed by a spring-loaded valve. There is an outlet, shown on the right of each diagram, that is connected to an actuator. The hollow, sliding piston provides an exhaust route to atmosphere.

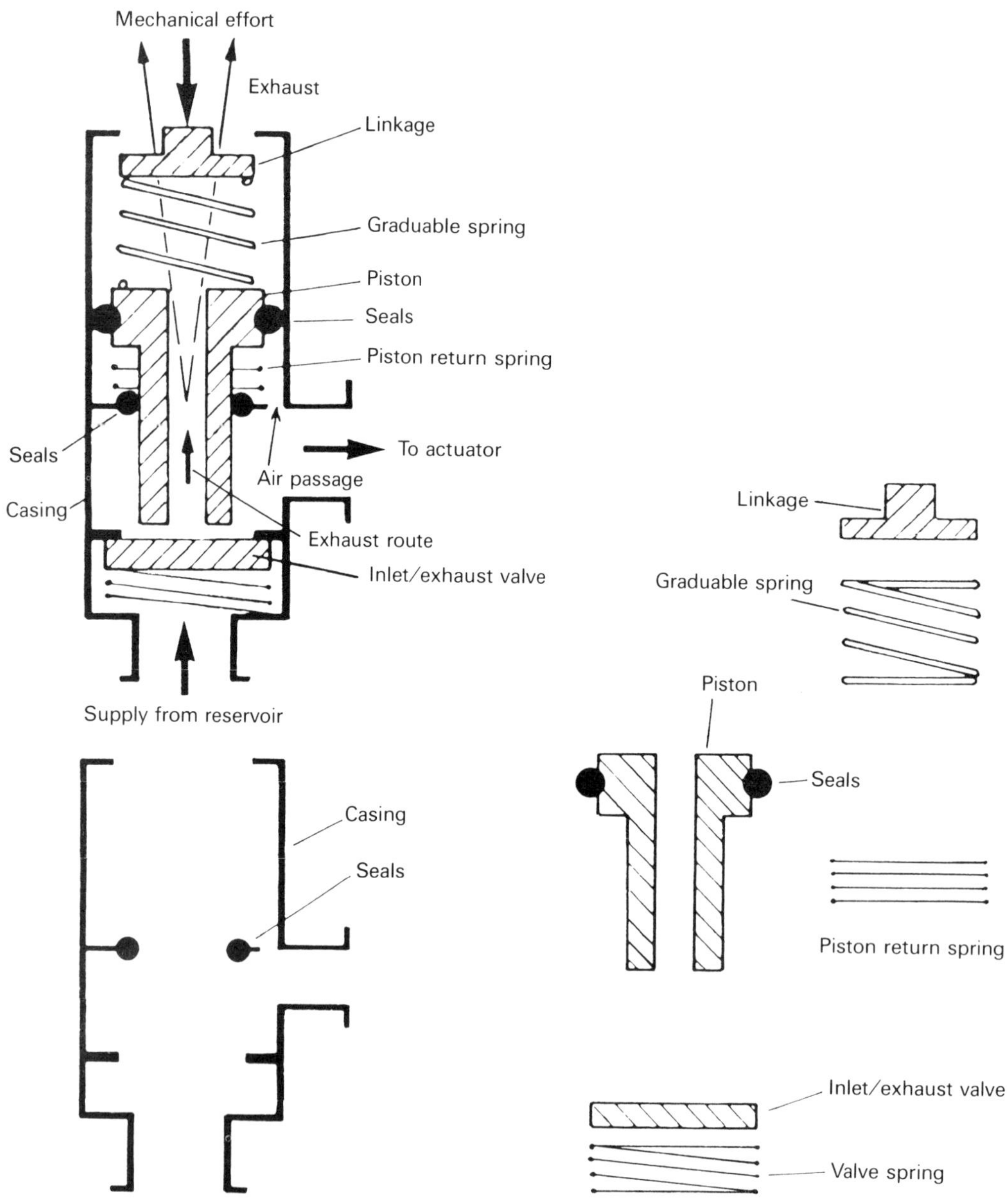

Figure 5.1 Components of a typical valve

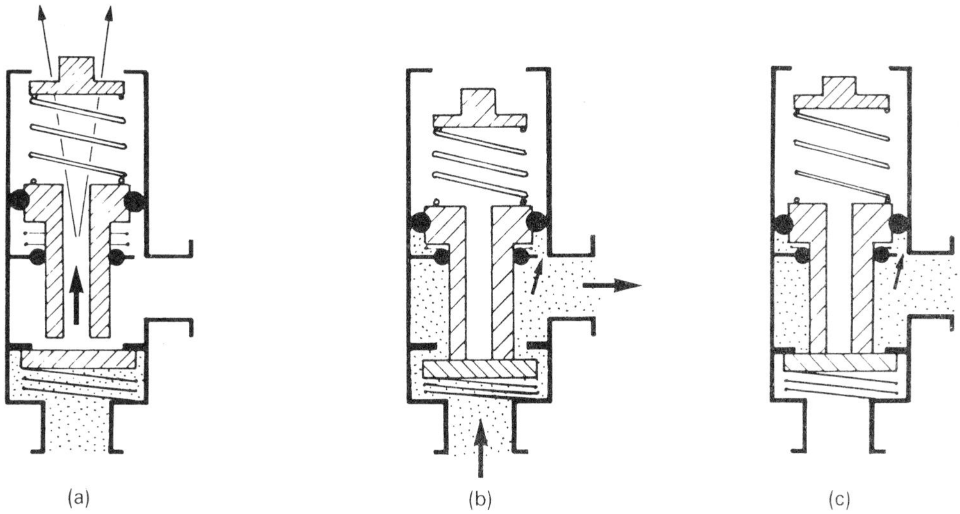

Figure 5.2 Valve action

Figure 5.2:

a) Shows the unit at rest. The supply is sealed. The line to the actuator is open to atmosphere via the hollow piston.

b) Shows that the hollow piston has been moved down to unseat the inlet/exhaust valve. Thus, the compressed air enters the unit, but as the lower end of the piston is now sealed, it can only leave through the port to the actuator.

c) Illustrates a position where the piston is still sealed against the inlet/exhaust valve, but this valve is also closed. This is the result of the pressure of the air below the shoulders of the piston exceeding the mechanical force applied above, so moving the piston slightly upwards. This is known as the *lapped* (or *balanced*) *condition*.

6 Dividing an air brake system into essential sections

The basic layout of air brake systems

All systems contain *four* sets of essential components:

1 compression/storage
2 controls
3 actuation
4 indication/warning.

When a vehicle that is fitted with compressed air brakes is in use, the following events occur:

1 Air is compressed and stored, ready for use.
2 When the driver operates either the foot or hand control, this causes air to be routed to the actuator required.
3 The actuator reacts to the compressed air supplied, applying the brakes mechanically.
4 Pressure gauges and low-pressure devices warn the driver of potential danger, enabling him to stop the vehicle safely, in the event of some defect developing. Brake stop-light switches are also air-operated.

Compression/storage

Atmospheric air is drawn into the compressor, via a filter, and the pressure is increased by compression to around 120 lbf/in^2 (827 kN/m^2, 8.3 bar). This compressed air is then passed to, and kept in, a storage tank (reservoir) ready for use. (Anti-freeze devices are used to prevent moisture freezing and rendering the system inoperative.)

However, consider the fact that a compressor is driven by the engine (or gearbox) through some form of transmission (for example gears or belt), which do not incorporate any form of clutch or means of temporary disengagement. If a vehicle is employed on normal road use, it is quite usual to find that part of a journey can be in congested city streets and part on open motorways. In the city, the brakes will be applied and released almost continuously, thus consuming large quantities of compressed air. However, on the motorway, the same vehicle could well travel 100 miles without the brakes being used and yet the compressor is still connected to the reservoir. Obviously some device is required to prevent dangerously high pressures building up, but which still allows the system to work when needed. There are two common ways of controlling the charging and maximum pressure in the reservoir:

1 By an unloader valve (see page 17) fitted in series between the compressor and reservoir.
2 By a governor valve (see page 38) fitted in parallel with the compressor–reservoir supply line and linked to an unloading mechanism on the compressor cylinder head, acting upon the inlet valve. Even when either of the above items are fitted, a safety valve must always be fitted in the circuit to allow for the failure or neglect of these components.

Controls

The components in this group draw air from the reservoir and supply it to the actuator, when demanded by the driver, in a controlled manner. There are two basic types of control valve:

1 **Progressive type** (often called graduable) Both the foot and hand valves are of this type, so that drivers can decide what degree of braking they require by the amount of force that they apply to the footpedal, or by the amount of movement applied to the hand lever of the valve.
2 **On/off switch** One common type is a toggle or flick switch, which is ideal for the differential lock or two-speed axle, where merely switching on or off is necessary. Another type is that used with an air-assisted handbrake. However, this type of handbrake cannot be classified as a secondary system, as legislation stipulates that a secondary system must have a progressive type of control.

Actuation

Compressed air, delivered via the control valve, is used to apply a force to a diaphragm or piston, which is linked to the brake shoes via mechanical linkage.

Indication/warning

For safety, and to comply with legislation, air brake systems are fitted with optical or acoustic warning devices, to give a signal to the driver when the air pressure in any part of the braking system falls to a point where it is still possible to apply *four* full applications of the service brake, and a *fifth* application at secondary performance.

Pressure gauges (see Figure 6.51 on page 44), on the dashboard indicate the air pressure in storage reservoirs. Low-pressure indicators (see Figure 6.52 on page 44) are electrical switches, with spring-loaded contact points held apart by a certain amount of air pressure. When the air pressure is low, the return spring will close the points and complete the electrical circuit. Some vehicles are fitted with a warning buzzer, others have a warning lamp and some have both devices.

The stop lamps on road vehicles fitted with air brakes, are illuminated in response to the operation of a brake control by the driver. Some braking systems use upright air (normal compressed air) to close the contact points in the electrical circuit, and others use inverse air (reducing air pressure) and the return spring closes the contact points. Therefore, there are two different types of stop-light switch available.

Refinements

There are many different valves that serve their own particular purposes, but within the general framework of basic systems. A number of these are explained in the following pages, but it is not practical to cover every type and it is suggested that the manufacturer of the system should be consulted if doubts arise regarding the function or functioning of any component.

A simple air brake system

This type of system is shown in Figure 6.1. A compressor supplies air to a storage reservoir via an unloader valve. The foot valve is supplied with compressed air from the reservoir and the force applied by the driver's foot controls the pressure of the air that is allowed to reach the brake chamber. The diaphragm within the brake chamber moves the push-rod assembly and from this point the operation is by mechanical action.

Compressors

These can be single or twin-cylinder, air or water cooled; Figure 6.2 shows a single-cylinder compressor. The design is very similar to a motor cycle engine, except that it lacks fuel and ignition systems and the valves are spring-loaded discs instead of the poppet

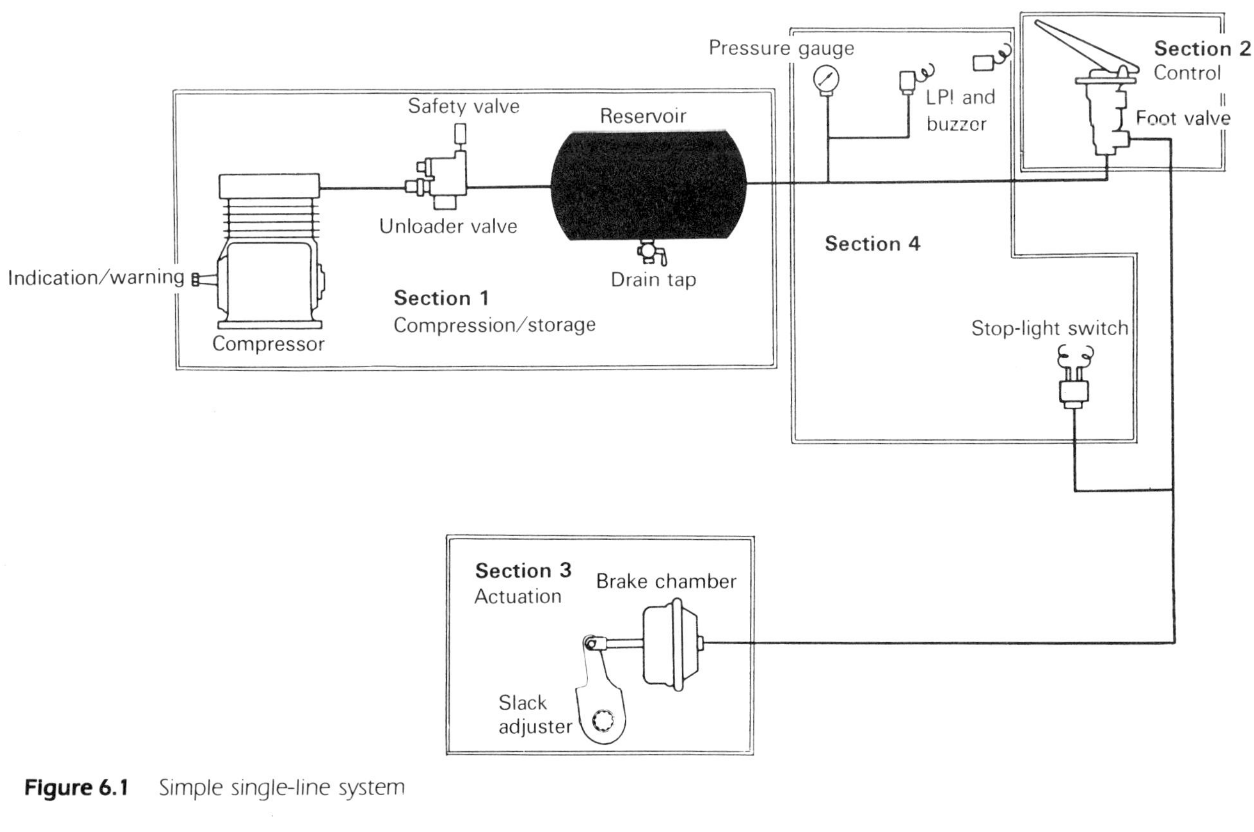

Figure 6.1 Simple single-line system

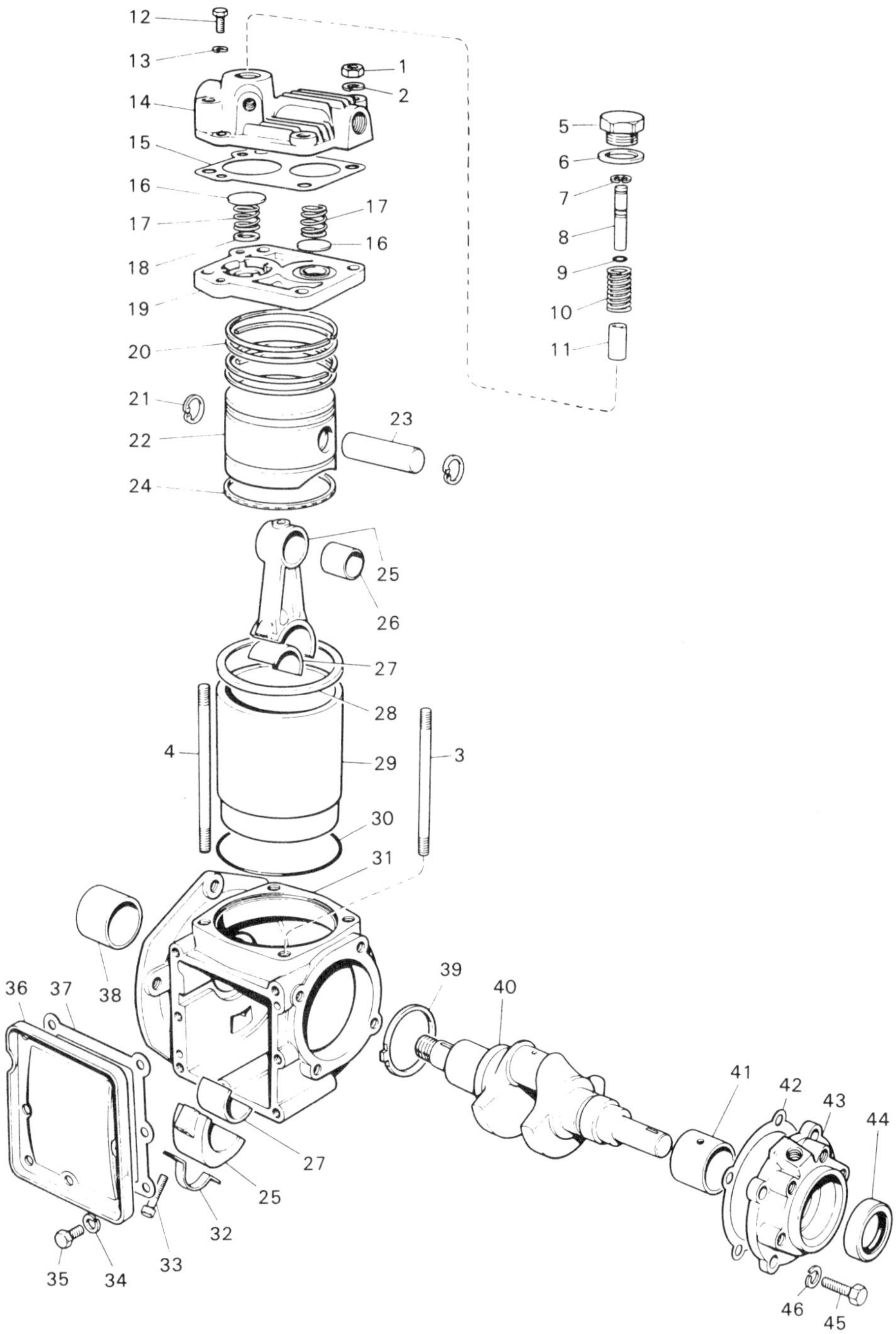

Key
1 = nut
2 = washer
3 = stud - short
4 = stud - long
5 = cap
6 = washer
7 = circlip
8 = unloader plunger
9 = sealing ring
10 = spring
11 = bush
12 = setscrew
13 = washer
14 = cylinder head
15 = joint
16 = valve
17 = valve spring
18 = shim
19 = baseplate
20 = piston ring
21 = circlip
22 = piston
23 = gudgeon pin
24 = oil scraper ring
25 = connecting rod
26 = bush
27 = large end bearing
28 = sealing ring
29 = cylinder
30 = sealing ring
31 = crankcase
32 = locking tab
33 = large end bolt
34 = washer
35 = setscrew
36 = plate
37 = joint
38 = bush
39 = thrust washer
40 = crankshaft
41 = bush
42 = joint
43 = cover
44 = oil seal
45 = setscrew
46 = washer

Figure 6.2 Single-cylinder compressor

type. Some types use reed valves, which are sections of spring steel fastened at one end, to assist air flow. Even a mechanical clutch, triggered by compressed air, has been used to allow a compressor to run light, without the aid of an unloader valve or governor valve. A through drive arrangement is a safety device where the fuel injection pump is driven via the compressor, hence if the compressor's crankshaft ceases to turn, the injection pump will not work and the vehicle's engine will stop.

There are approximately 400 types of compressor in current use, and therefore, when specifications of particular models are required they should be obtained from the manufacturer.

General operation

As the piston descends, the partial vacuum created above unseats the inlet valve, allowing air to enter the upper cylinder. As the piston rises, the compressed air, aided by the spring on the inlet valve, closes the inlet valve. As the piston continues to rise the pressure rises lifting the delivery valve (equivalent of exhaust valve) and allowing the now compressed air to pass through the line, en route to the reservoir. As the piston again descends, the delivery valve reseats to prevent the departing air returning, and the whole cycle is repeated.

Winter operation

Compressed air supplied by a compressor contains a proportion of water vapour; the percentage is dependent upon the humidity of the area in which the vehicle is operating. Consider compressing a large volume of atmospheric air into a small reservoir. Although the air is then compressed it still contains the same amount of moisture that was contained in the much larger volume.

In conditions where temperatures are low, water (condensed vapour) will freeze, damaging components and/or causing a total seizure of control valves and operating chambers, thereby causing a total brake failure. One method of preventing such an occurrence is by the introduction of alcohol into the compressed air within the braking system, but a more satisfactory and permanent arrangement is achieved by fitting an air dryer.

Air filter plus tyre inflator (WO)

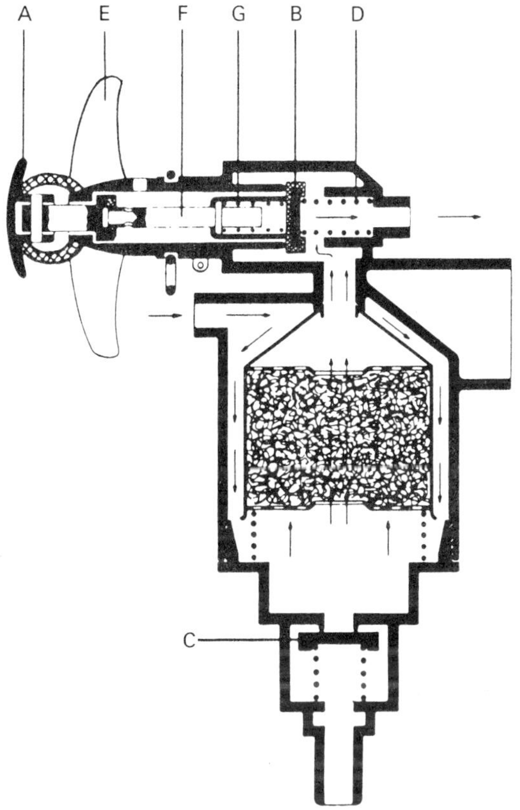

Figure 6.3 Air filter and tyre inflator (WO)

As shown in Figure 6.3, air from the compressor enters the port at the top of the canister, before descending between the double-skinned casing and then rising again, through the filter element, en route to the unloader valve. Dirt and oil that accumulate above relief valve (C), can be ejected manually, by pressing knob (A) onto its stop, whilst the engine is running. Such action causes disc valve (B) to seal off the route to the unloader valve and thus the compressor output is solely charging this small canister.

Pressure will rise rapidly and at approximately 162 lbf/in^2 (1116 kN/m^2, 11 bar), relief valve (C) will open and air pressure will eject the accumulated dirt and oil. When knob (A) is released, return spring (D) reseats disc valve (B) and the normal charging of the unloader valve circuit recommences.

Before inflating tyres, it is important that this unit should be drained. Then, after unscrewing knob (A) and wing nut (E), the inflator hose can be affixed. When the hose is screwed onto this unit, stem (F) will cause spring (G) to unseat disc valve (B), thus closing off the route to the unloader valve, so that the compressed air is directed to the tyre.

When the hose is removed, spring (D) again reseats disc valve (B), thus reopening the unloader valve circuit.

Line filter (WO)

Although the type of filter described above (if fitted) is to be found between the compressor and the unloader valve, another type of filter (a line-type) can be fitted in other parts of the system and even on a trailer. These units are designed to prevent dirt entering valves, which naturally can cause deterioration, excessive wear, malfunction and/or failure.

As shown in Figure 6.4, compressed air enters port (2), passes through filter (A) and leaves by port (1).

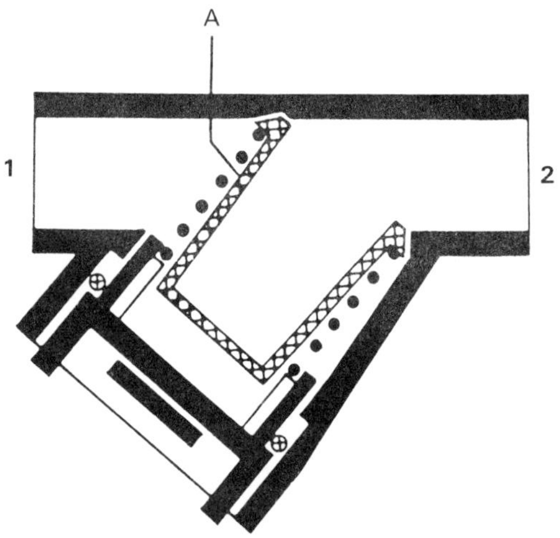

Figure 6.4 Line filter (WO)

However, as with lubrication systems, filters can be forgotten (whether by neglect or mistake) and a complete system must not be allowed to be starved merely because the filter has become blocked. Therefore, the designers have worked on the same theory as those who produce lubrication systems, i.e. in the event of a filter blockage it is better to receive contaminated supplies than no supplies at all. As with lubrication systems, the filter element is spring-loaded, so that if blocked, pressure moves the element off its seat to allow flow to continue.

Air silencer (WO)

This device has been developed to reduce the noise caused by compressed air leaving the exhaust port of a valve. As shown in Figure 6.5, compressed air enters port (1), and passes through filter (A) to reach airway (B). The air speed is thus reduced, dampening the noise. Air pressure unseats diaphragm (C), opening exhaust port (3) to allow the compressed air to escape to atmosphere.

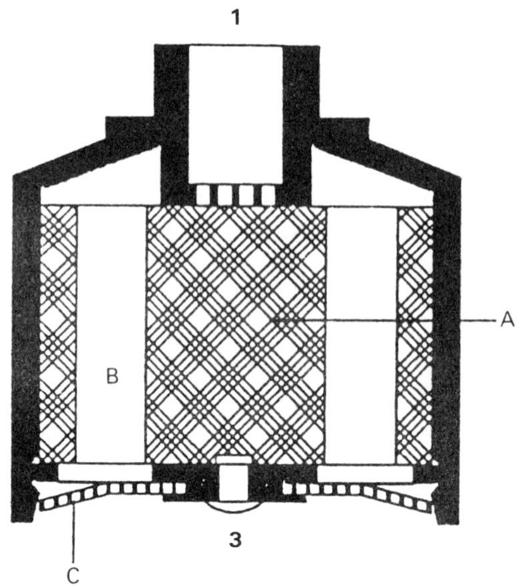

Figure 6.5 Air silencer (WO)

Alcohol evaporator and check valve (BW)

The alcohol evaporator (see Figure 6.6) allows vapourised alcohol to enter the compressed air system to lower the freezing point of the moisture (water vapour) that has been drawn in from the atmosphere, and thus reduce the risk of damage or failure of equipment in low temperatures.

An alcohol evaporator is connected to the inlet manifold of the compressor. When the compressor is in operation, a partial vacuum is created in the inlet manifold which is transmitted to the volume within the plastic bottle (C), above the level of the alcohol. As a result of this drop in pressure within the bottle, atmospheric air enters through filter (E), passing through the evaporator tube (D) and emerging at the bottom of the bottle. This incoming air travels up through the alcohol, creating a vapour as it breaks the surface, and this resulting vapour passes out through pipeline (A) and into the manifold of the compressor.

The plastic bottle should only be filled with alcohol when protection from frost is actually required. An additional safeguard against waste is an alcohol evaporator check valve (see Figures 6.7 or 6.8, depending upon the system) which limits the evaporation of alcohol by sealing off the evaporator, when the compressor is not delivering compressed air to the reservoir. The type of check valve fitted to the evaporator depends upon whether the working pressure of the particular braking system is controlled by an unloader valve or a governor valve.

Alcohol

The best results can be obtained by using methyl alcohol (CH_3OH) which is marketed under the names of *Blending Methanol* or *IMS Substitute*.

Caution: These substances are both highly inflammable and toxic, whether in liquid or vapour forms. Any exhaust from control valves in the cab of a vehicle should not discharge into the interior of the cab.

Proprietary anti-freeze mixtures, as used in cooling systems, will not work in this type of anti-freeze operation and should not be used.

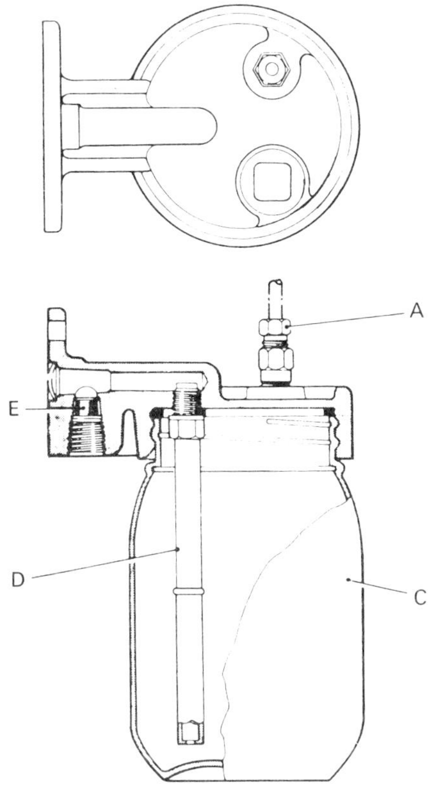

Figure 6.6 Alcohol evaporator (BW)

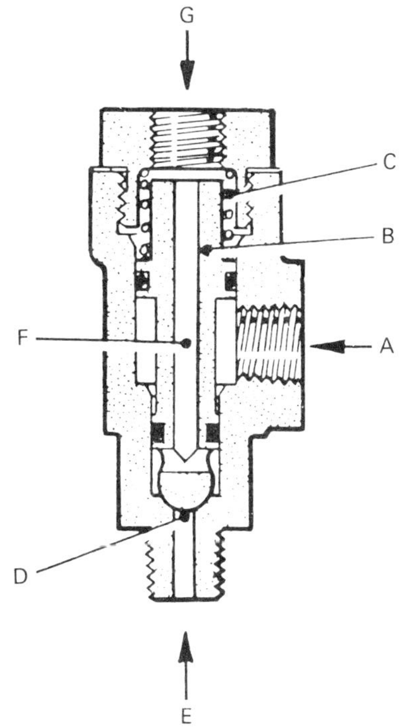

Figure 6.7 Alcohol evaporator unloader system check-valve

Check valve for an unloader system

As shown in Figure 6.7, whilst the compressor is delivering compressed air to the reservoirs, a signal pressure is delivered from the pipeline between the compressor and unloader valve to port (A) on the check valve. This pressure moves piston (B) upwards against the resistance of spring (C), opening the valve (D) at the seat. Thus a route is provided from inlet port (E), connected directly to the evaporator, through central drilling (F) in the piston, to outlet port (G) which is connected to the inlet manifold of the compressor. When the design reservoir pressure has been achieved and the unloader valve is discharging to atmosphere, the compressor will 'run light' and the signal pressure to port (A) is reduced. Thus spring (C) reseats piston (B) at seat (D), stopping further supplies of alcohol.

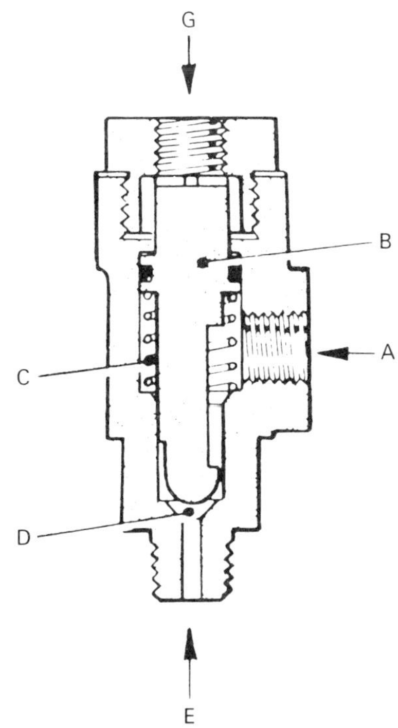

Figure 6.8 Alcohol evaporator governor system check-valve

Check valve for a governor system

As shown in Figure 6.8, a pipeline, which is connected between the governor valve and the unloader mechanism of the compressor, is connected to port (G). Whilst the reservoirs are being charged, and air pressure has *not* reached governor cut-out pressure, compressed air is *not* present at port (G) of the check valve. At this stage piston (B) is held off seat (D) by spring (C).

A partial vacuum, caused by the pumping action of the compressor, is relayed by a pipe to port (A) of the check valve, causing alcohol vapour to be drawn from the evaporator and passed to the braking systems.

When governor cut-out pressure is reached, compressed air passes to the unloader mechanism of the compressor and, additionally, to port (G) of the check valve, causing piston (B) to move down, against spring (C), thus closing port (E). This prevents further alcohol entering the braking systems until governor cut-in pressure is again reached.

Figures 6.9 and 6.10 show the piping diagrams for an alcohol evaporator in an unloader system and a governor system respectively.

Alcohol injector

This is a device which gives positive delivery of a calibrated amount of alcohol into the compressed air system, beyond the compressor.

It can be seen, in Figure 6.12, that there is a pipework connection between the delivery port of the injector and the pipework linking the compressor to the wet tank. Another pipe connection to the injector links with the pipework between the governor and the unloader mechanism of the compressor. (An alcohol injector can also be used in an unloader valve system.)

As shown in Figure 6.11, when the pressure in the reservoir rises to the design working pressure, the governor sends a signal pressure to the unloader mechanism, and this pressure also enters the injector

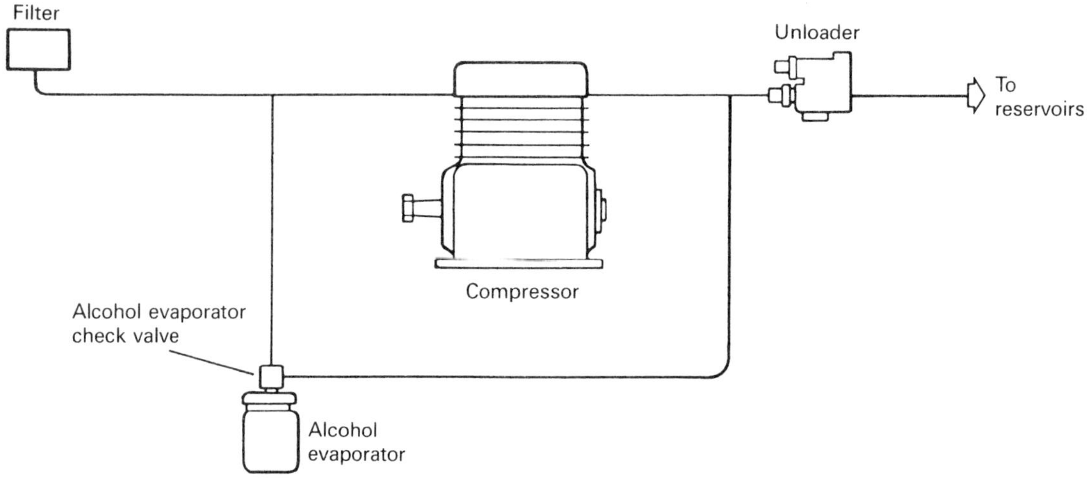

Figure 6.9 Alcohol evaporator – piping diagram (unloader system)

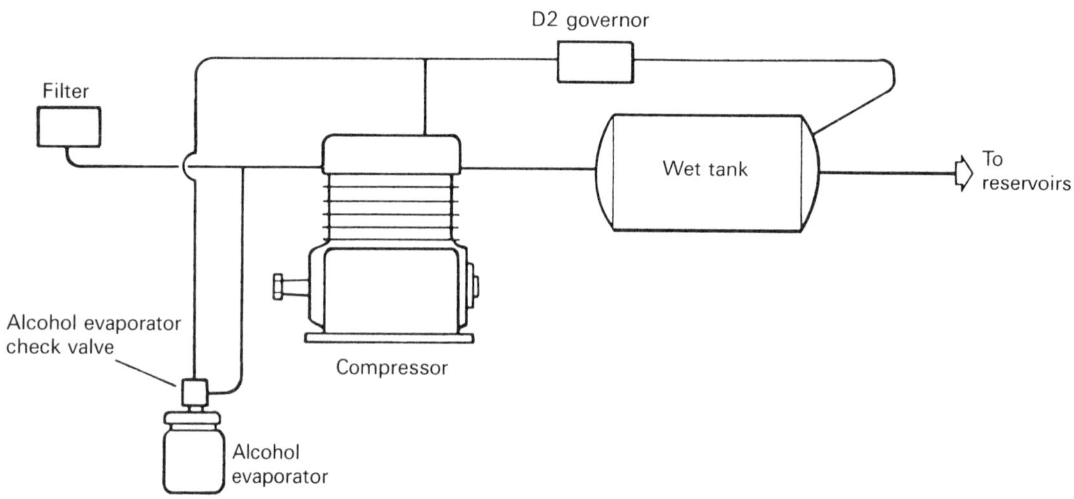

Figure 6.10 Alcohol evaporator – piping diagram (governor system)

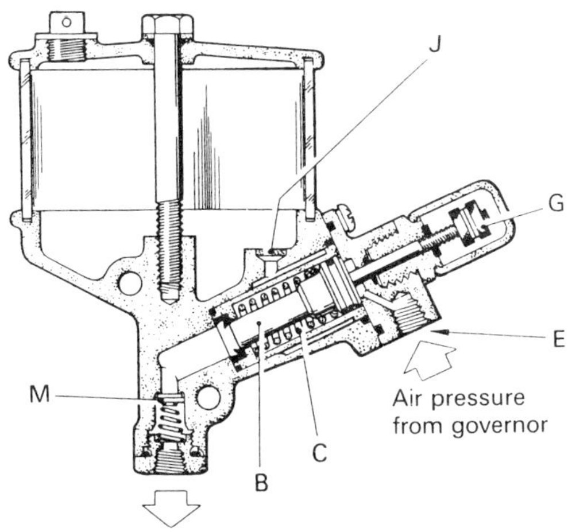

Figure 6.11 Alcohol injector (BW)

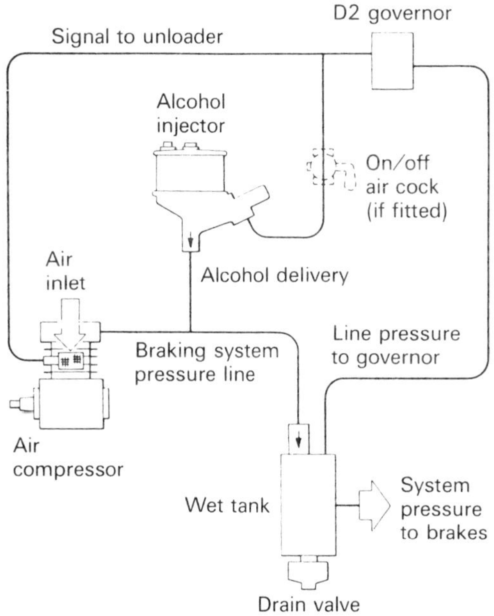

Figure 6.12 Alcohol injector - piping diagram

via port (E), causing pump plunger (B) to inject alcohol into the compressed air system. When the pressure at the governor drops, the signal pressure ceases and the pump plunger in the injector is returned by the pressure of spring (C). On the return stroke of the plunger, fresh alcohol from the reservoir is drawn into the pump barrel through filter (J). A non-return valve (M) prevents loss of alcohol. Adjustment of quantity of alcohol being delivered can be adjusted by two nuts (G).

In warm weather this unit can be rendered inoperable by screwing down the adjuster nuts to lock the plunger, or by the use of an on/off cock (if fitted).

Anti-freezer pump (WO)

This is an anti-freezer pump which has to be manually-operated several times before commencing a journey (in severe weather), and needs correspondingly more operations if a trailer is being drawn, due to the extra air consumption involved. After such a process, several brake applications must be made to ensure that the anti-freeze solution has been spread throughout the whole of the brake valves and pipework on the vehicle/trailer.

As shown in Figure 6.13, feed holes link the reservoir with the pump cylinder. As the operator depresses the plunger rod, the piston is forced down, sealing off the feed holes, and causing approximately 1 cm^3 of anti-freeze solution to be injected, via the check valve, into the compressed air passing through the base of this unit, en route from the compressor to the vehicle's reservoirs.

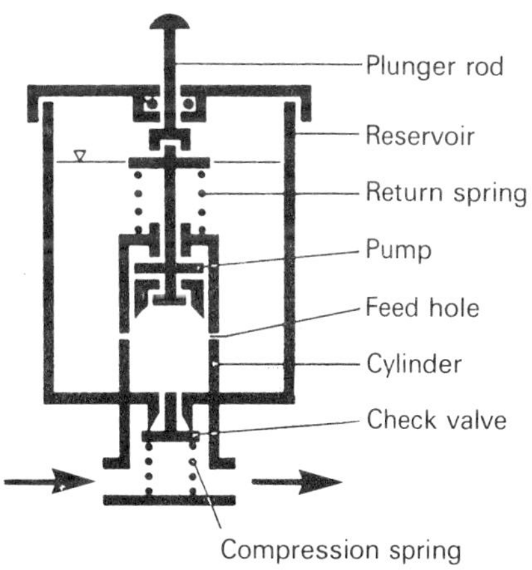

Figure 6.13 Anti-freezer pump (WO)

Whilst the plunger rod is held down, further anti-freeze solution cannot flow from the reservoir. When released, the plunger and piston are returned to the 'off' position, by the return spring. Similarly, the check valve is closed by the return spring. Feed holes allow anti-freeze solution to feed the cylinder, and the pump is again ready for use.

Anti-freezer (KR)

As shown in Figure 6.14, when the storage section of the air brake system has reached its correct working

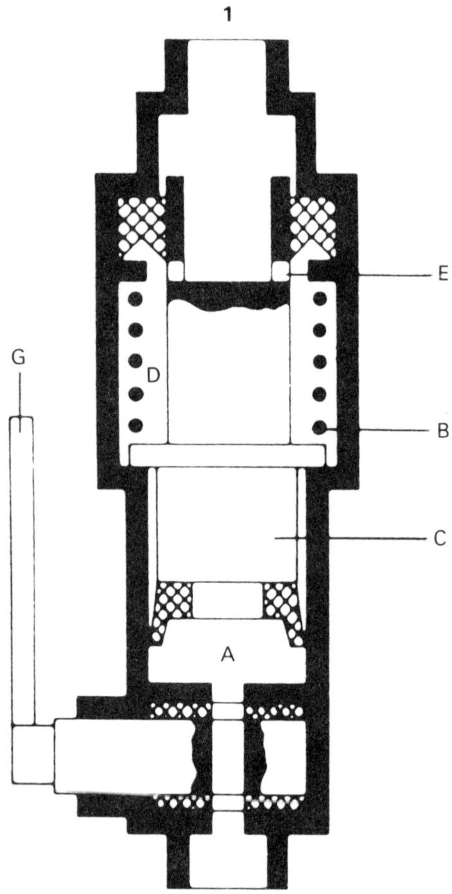

Figure 6.14 Anti-freezer (KR)

pressure and the unloader valve is discharging to atmosphere, pressure at port (2) and chamber (A) drops to zero. Spring (B) ensures that piston (C) remains in the closed position. Alcohol (from a separate reservoir) enters chamber (D) via port (1) and cross drillings (E).

When the unloader valve ceases to discharge to atmosphere, and restarts to feed the storage reservoirs with compressed air, pressure rises at port (2) and chamber (A), lifting piston (C) against spring (B). This closes cross-drillings (E), and a small quantity of alcohol passes from chamber (D) into chamber (A) and out of port (2) into the air flow to the storage reservoirs of the brake system. (This unit can be isolated in warm weather by using lever (G).)

Anti-freezer (WO)

This unit (see Figure 6.15) is fitted into the pipeline between the output of the compressor and the input of reservoirs. It relies upon compressed air flowing over an alcohol-soaked wick, which protrudes through an inlet port. The wick draws alcohol from its own reservoir. In warm weather the wick can be lowered, out of the air flow, by merely pushing down the toggle handle and turning it through 90° to lock it in the 'off' position.

Anti-freeze pump (WO)

The role of this anti-freeze pump (as shown in Figures 6.16A and 6.16B is to inject a measured amount of

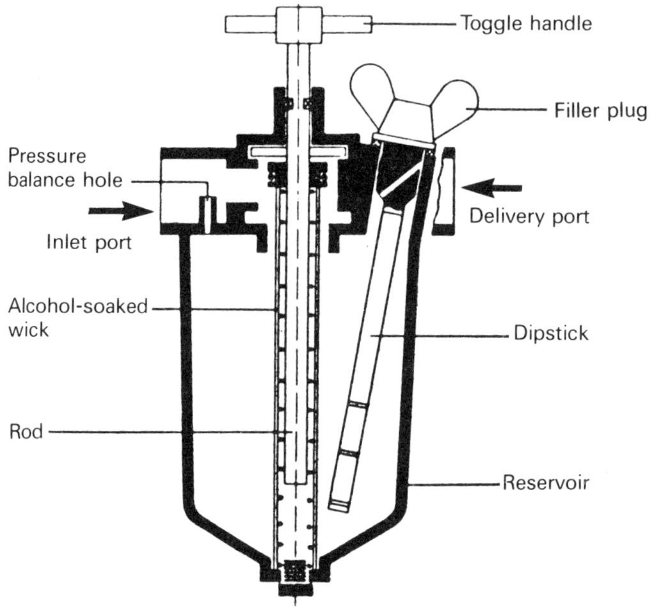

Figure 6.15 Anti-freezer (WO)

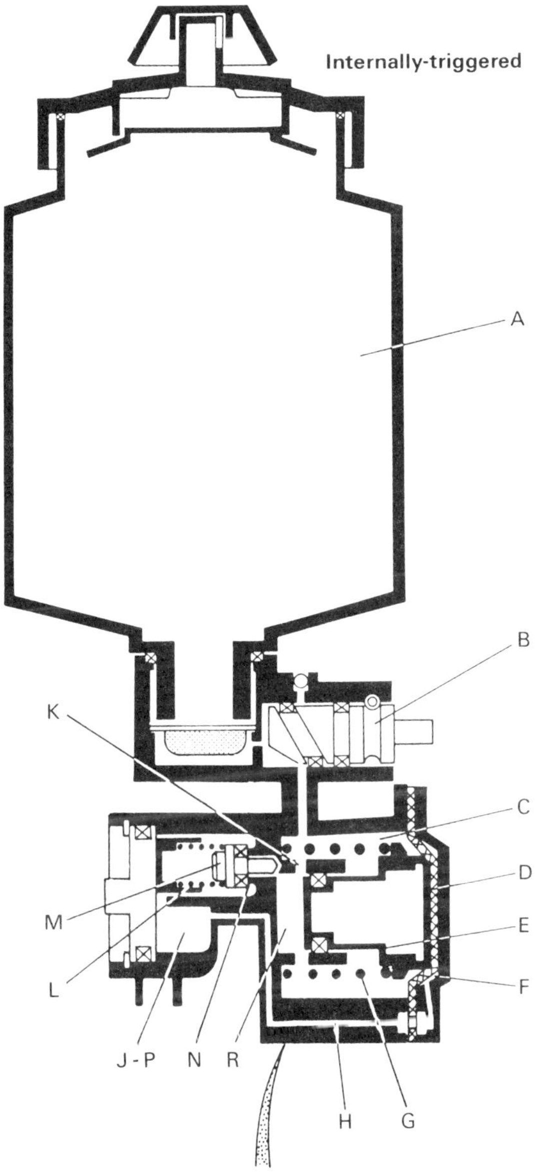

Figure 6.16A Anti-freeze pump - internally-triggered type (WO)

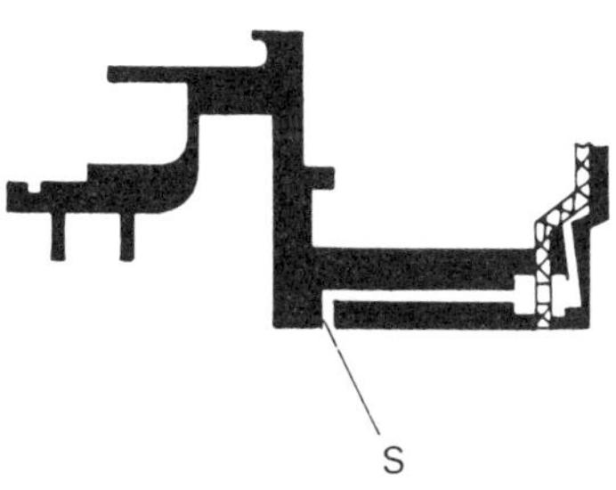

Figure 6.16B Anti-freeze pump - externally-triggered type (WO)

anti-freeze liquid into the compressed air of the braking system, to prevent any moisture from freezing, when lower temperatures are encountered. During warm weather, the reservoir can be isolated by the use of turn-cock (B).

Internally-triggered type

As shown in Figure 6.16A, this unit is located in the air delivery line, between the compressor and the unloader valve. Air can flow in either direction through ports (J) and (P).

Plunger (E) is activated by air pressure in the delivery line, which passes through drilling (H), to reach chamber (F) and which is at a sufficiently high pressure to cause diaphragm (D) to move plunger (E) to the left, against its return spring (G).

Anti-freeze liquid from reservoir (A) is gravity fed into chamber (C) and through drilling (K), to reach injection chamber (R). When plunger (E) has moved to the left and passed drilling (K), anti-freeze liquid in chamber (R) is pressurised and acts upon the underside of valve (M).

When this pressure overcomes line pressure at (J-P), spring (L) will move valve (M) off seat (N), and the volume of anti-freeze liquid trapped in (R) will be injected into the delivery line, via open valve seat (N), and thus throughout the braking system.

When the unloader valve in the braking system cuts-out, that valve discharges the compressor output to atmosphere. Consequently, there will also be a pressure drop in the delivery line of the anti-freezer pump. This pressure drop affects chamber (F) and, therefore, the plunger return spring (G), overcomes the air pressure in chamber (F), reseats diaphragm (D) and plunger (E). As plunger (E) uncovers drilling (K), chamber (R) is recharged with anti-freeze liquid, from chamber (C). The pump is ready to inject again, when the unloader valve next cuts-out.

Externally-triggered type

As shown in Figure 6.16B the pump can be installed in the line between the unloader valve and the reservoirs (unloader system), or between the compressor and the reservoirs (governor system). Anti-freeze injection is the same as for the internally-triggered type, but the signal to operate is received from the unloader or governor valve. The signal enters chamber (F) through port (S), to cause the injection cycle described for the internally-triggered type.

When the unloader valve or governor valve cuts-in,

that drop in pressure is communicated to chamber (F) and as a result the recharging of chamber (R) with anti-freeze liquid occurs.

Air dryer (BW)

Atmospheric air contains water, in the form of vapour, which together with dirt and oil cause deterioration of many components within an air brake system. Designers are always striving to reduce the amount of moisture that is allowed to reach the valves. Wet tanks, condensers, drain cocks and automatic drain valves are all in common use. The air dryer is a component aimed at removing the three contaminative items from the system. It is connected in series between the compressor and the reservoir.

As shown in Figure 6.17, air from the compressor enters port (F) and flows downwards between the dryer body and the outside of cartridge container (G). The air is cooled by contact with the cool metal and the water/oil that is condensed from the air and collects in the bottom of the dryer, as does most of the carbon and dirt. The now cool air travels upwards through oil filter (H) which extracts any remaining droplets of water/oil and particles of carbon/dirt.

As the air continues to rise it passes through the inside of cartridge (G) which is packed with numerous micro-crystalline desiccant (drying) pellets, which combined give an enormous surface area. These pellets absorb any remaining moisture as the air rises, until it passes through the check valve (L) in the desiccant plate (K). At this stage the air will be sufficiently dry that a drop in temperature of 30°F (17°C) would be necessary to cause any further condensation. The area above the desiccant plate is known as the purge chamber. Air passing through this section leaves through check valve (R), passing out of outlet (Q) and then to the reservoir.

Continual absorption of moisture would saturate the desiccant pellets, rendering them useless. Therefore, a reverse flow of dry air takes place whenever the reservoir pressure activates the governor valve, to send a signal pressure to operate the unloader mechanism on the compressor. By having a T-piece connection in the pipeline between the governor and the compressor, this signal is also delivered to port (C) on the dryer. Compressed air entering port (C) opens purge valve (A) which permits the existing air pressure within the dryer to expel any liquid and dirt that may have accumulated. This loss of pressure in the dryer causes the purge chamber to discharge dry air through the purge vent (M) into the desiccant cartridge. This compressed air, expanding back to atmospheric pressure, takes moisture from the desiccant as it escapes to the atmosphere via the open purge valve.

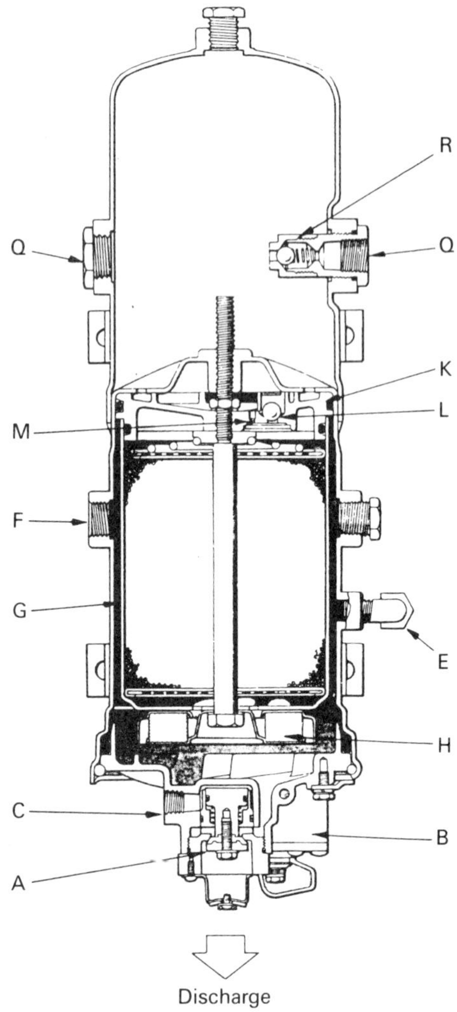

Figure 6.17 *Air dryer (BW)*

Should the compressor continue to unload after the purge chamber has been exhausted, the check valve (R) will prevent any loss of pressure in the reservoir supply line. When the compressor commences to recharge, the whole cycle is repeated.

There is a 60 watt electric heater (B) connected to the master electrical switch (ignition switch in the case of SI engines) of the vehicle, and linked to a thermostat that cuts in at 50°F (10°C) and cuts out at 85°F (29°C). This arrangement prevents the condensate from freezing in cold climates.

A safety valve (E) is fitted to this unit.

Twin-box air dryer (WO)

This unit basically consists of two vertically-mounted canisters of desiccant, which each have oil/dirt filters at both top and bottom. It is fitted in systems using an unloader valve and is capable of operating in temperatures between −40 °C and +65 °C.

As shown in Figure 6.18, air from the compressor enters port (1) and passes through gauze filter (m), passage (E) and valve (l) to reach the area between outer casing (A) and cannister (d). It then passes through felt filter (b), desiccant (c) and felt filter (f) into passage (G).

The, then dry, air divides between *three* routes. By one of these routes, this air passes through non-return valve (g) and leaves the air dryer through port (2), en route to the storage reservoirs of this vehicle. By a second route, the dry air flows from passage (G), into passage (D) to act upon the left-hand side of valve (p), thereby causing valve (o) to close the route between passages (E) and (H). The third route is through passage (G), the choke of disc valve (n) and (q), and back-flushing desiccant (r) from bottom to top, whereby the dry air collects the moisture from these granules within cannister (s). This air, now damp, descends between outer casing (K), cannister (s) and passage (H) to reach exhaust port (3).

At one-minute intervals, a time switch operates to cut off the supply of current to solenoid valve (h),

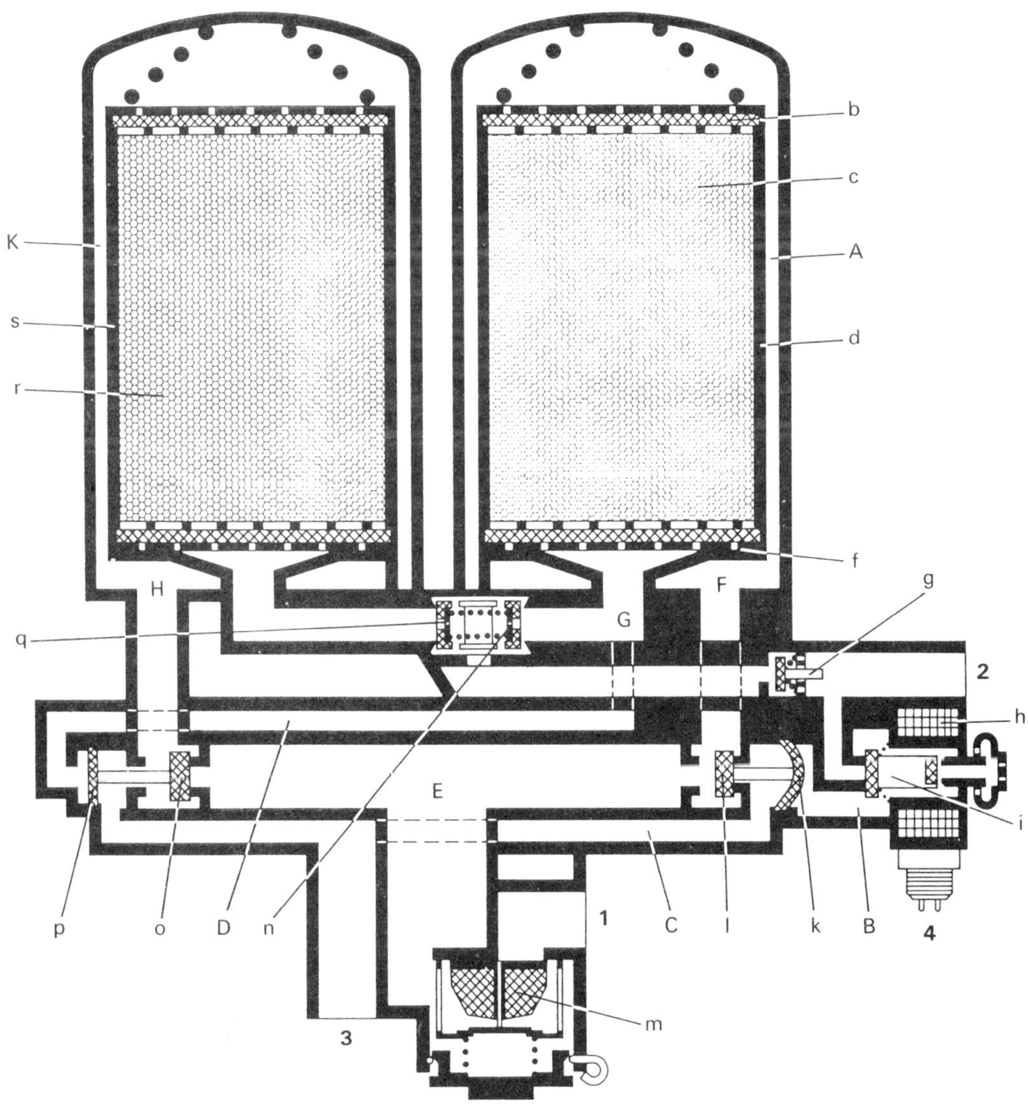

Figure 6.18 Twin-box air dryer (WO)

causing plunger (i) to retract to the right, thus allowing the air to flow from port (2), through passage (B), to act upon diaphragm (k).

Valve (l) closes the connection between passages (E) and (F). Air in passages (F), (G) and (D) escapes to atmosphere via passage (C) and exhaust port (3). A drop in pressure causes non-return valve (g) and disc valve (n) to close.

With a pressure drop in passage (D) and at diaphragm (p), the pressure in passage (E) opens valve (o), which allows air from the compressor to enter chamber (H). This air then passes between outer casing (K) and cannister (s), and enters through the top filter of that cannister to pass through the desiccant which absorbs the moisture that the air contains.

The ejection of the moisture, by compressed air, and the drying cycle are then repeated at one-minute intervals.

Air dryer (RR)

Anchorlok air dryers are produced in two forms; vertical and horizontal models.

The vertical type can be added to an existing air brake system, while the horizontal version is intended as a replacement for a wet tank in an existing system.

This air dryer is a little more sophisticated than a simple wet tank, as it has a double skin, with external cooling fins.

Incoming hot air, from the compressor, passes between the two surfaces, and its temperature is rapidly reduced, causing water/oil vapour to condense into liquid, which will drop to the base of the unit and be ejected by an automatic drain valve.

This unit contains neither a filter nor a dessicant.

Single cartridge air dryer (GN)

As shown in Figure 6.19, this component contains *five* items as follows:

1 A filter which extracts oil/dirt from the compressed air entering the dryer.
2 A cartridge containing desiccant.
3 Desiccant (moisture absorbing granules).
4 An automatic drain valve, which the manufacturers call an integral unloader valve.
5 A heating element, with thermostat control, to prevent freezing.

Figure 6.20 shows the route of the compressed air supplied by the compressor, through the air dryer, two non-return valves, and a quadruple protection valve, en route to the storage reservoirs. Simultaneously, compressed air is fed, via a T-piece, to the reservoir (5).

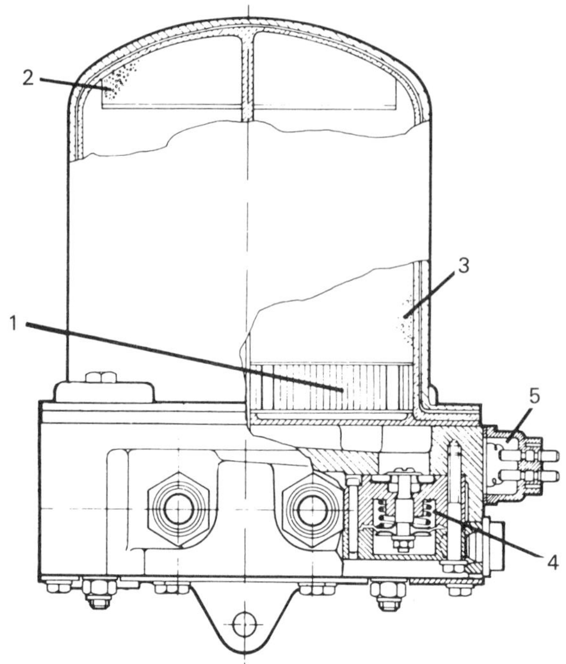

Figure 6.19 Single cartridge air dryer (GN)

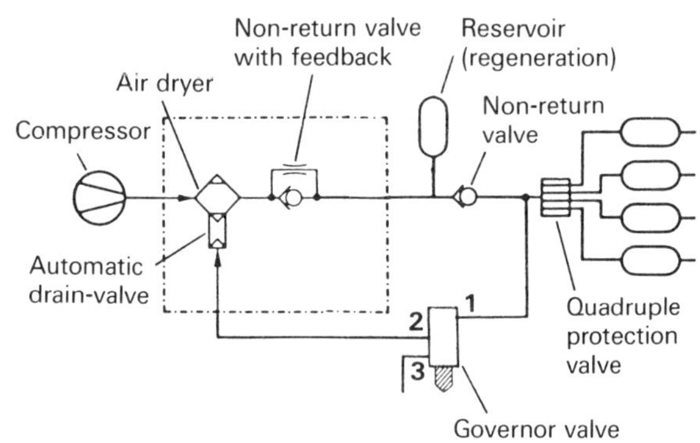

Figure 6.20 Piping layout for single cartridge air dryer

A governor valve is connected, in parallel, with the feed line to the storage reservoirs. When the air in these reservoirs reaches the recommended system pressure, the governor valve allows air to pass through port (2), to act upon and open the automatic drain valve on the air dryer. With this valve open, compressed air from the reservoir (5) feeds back through the non-return valve (4) with feedback, reverse flushing and drying the desiccant. The condensate extracted from the desiccant is expelled to atmosphere through the automatic drain valve of this unit.

This cycle will continue until the system pressure

drops, and the governor valve closes the through route to the automatic drain valve on the air dryer. Compressed air, then trapped in the pipeline between the governor valve and the automatic drain valve, will escape to atmosphere via port (3).

Condenser and drain valve (CD)

An alternative to an air dryer, is a condenser (fitted with an automatic drain valve), as shown in Figure 6.21. The purpose of this component is to condense the water and oil vapour in the air being supplied to the system by the compressor. This liquid is then expelled into the atmosphere, before it can reach other components, via the auto-drain that is fitted to the base of the condenser.

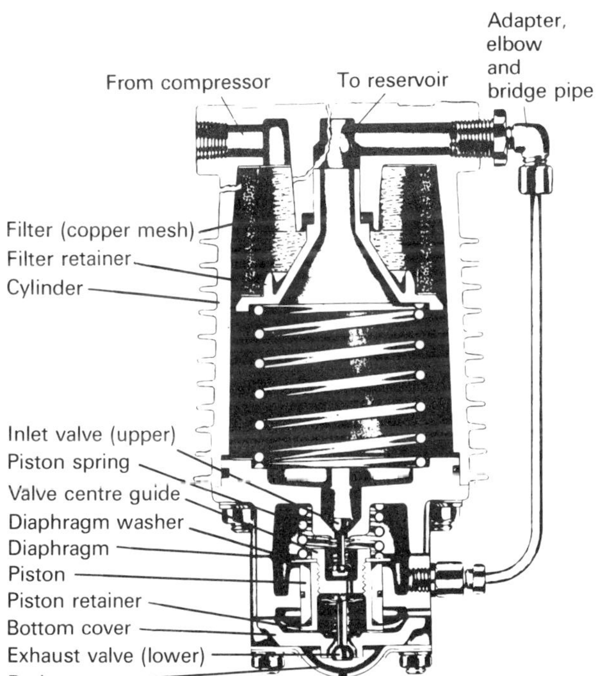

Figure 6.21 Condenser (CD)

A condenser can be fitted to any form of compressed air braking system, whether it has an unloader valve or a governor valve, but the pipe connections are different.

As shown in Figure 6.22, the condenser is connected in series between the compressor and the unloader valve. Compressed air from the compressor enters the condenser through its inlet port and then passes through the filter, which extracts dirt and oil droplets. The exterior of the cylinder has cooling fins so that the inner walls of the cylinder are cool for the hot incoming air to pass over. The action of the hot air impinging on

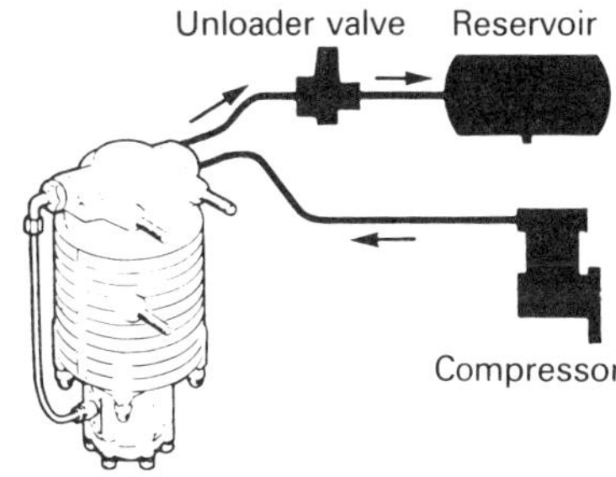

Figure 6.22 Condenser piping diagram – unloader valve system

the cold metal condenses vapour into liquid which gravitates to the bottom of the cylinder, whilst the air leaves by the delivery port.

If the filter becomes blocked, the air pressure will build up at the inlet port sufficiently to compress the filter retainer spring, so that the filter and its retainer move down the cylinder, allowing the air to bypass the filter. Thus, the air supplied to the whole system will be unfiltered.

When the air leaves the delivery port, en route to the unloader valve, it also feeds the bridge pipe and reaches the chamber beneath the upper diaphragm. An increase in pressure in the chamber lifts the upper diaphragm and piston, which pushes open the inlet valve. This action causes the exhaust, below, to close. The passage between the cylinder and drain valve is pressurised and the condensed liquid is forced down into the bottom cover.

When cut-out pressure is reached at the unloader valve, it discharges to atmosphere allowing the pressure to drop in the cylinder and bridge pipe. When this happens the return spring reseats the inlet valve, opening the exhaust valve, and the air pressure that has been trapped ejects any liquid present.

As shown in Figure 6.23, the same principles apply in a governor valve system as in an unloaded valve

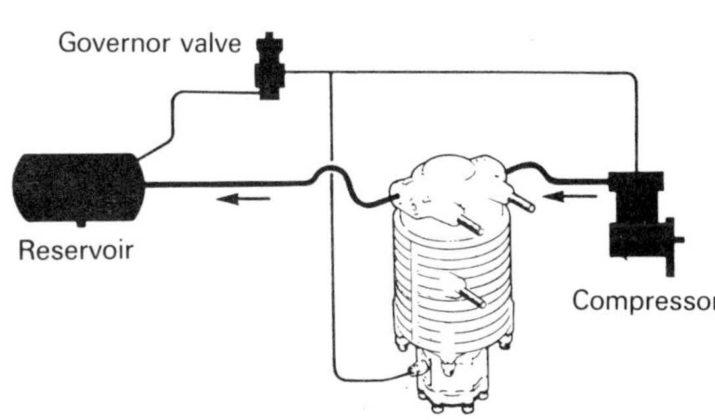

Figure 6.23 Condenser piping diagram – governor valve system

system, except that the pressure for opening the inlet valve is supplied from the feed-back pipe, between the governor and the unloader mechanism of the compressor.

Condenser with automatic drain valve (WO)

The role of this unit is to cool the air being delivered by the compressor to the reservoirs, and so condense moisture into liquid and then to expel this liquid and dirt to atmosphere, via an automatic drain valve.

As shown in Figure 6.24, compressed air, from the compressor, enters port (1), past a T-connection to a safety valve. This air then unseats non-return valve (a) and passes into chamber (A) en route to delivery port (21). It passes through filter (b) before being forced between deflector (c) and the cylinder wall (d), which is being cooled externally by the passing air.

Thus, the incoming compressed air is cooled by contact with the comparatively cool, cylinder wall, prior to passing to port (21). The resulting liquid (condensate) drops to the base of chamber (B).

From this point the condensate is ejected by the automatic drain valve. This valve uses the same operating principle as the DV2-type, i.e. it uses changing pressure within the condenser to cause its operation. The pressure in chamber (B) can also be sensed via port (22), for governor-controlled systems.

There is a filter (b) in the condenser and another filter (f) in the auto-drain valve, but air can bypass either of these in the case of any filter becoming blocked.

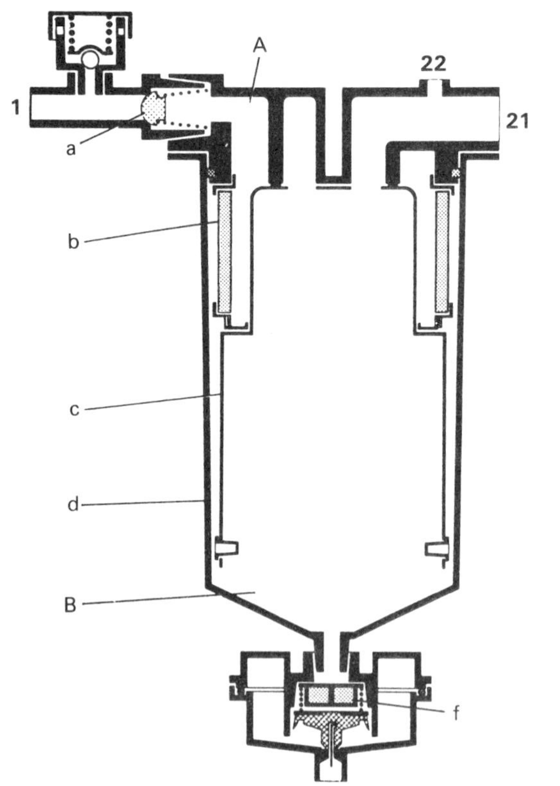

Figure 6.24 Condenser with automatic drain valve (WO)

Automatic drain valve (DV2) (BW)

This type of automatic drain valve, as shown in Figure 6.25, is only suitable for use with normal-size reservoirs and is not suitable for small-capacity tanks, such as wet tanks or sensing tanks with a capacity of less than 300 in^3 (4916 cm^3).

With the reservoir at atmospheric pressure, inlet valve (E) and exhaust valve (K) are against their seats. This is shown in Figure 6.26.

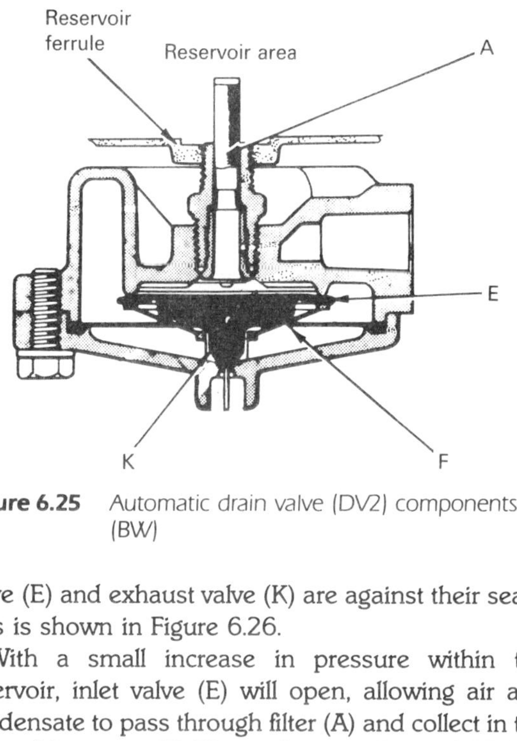

Figure 6.25 Automatic drain valve (DV2) components (BW)

With a small increase in pressure within the reservoir, inlet valve (E) will open, allowing air and condensate to pass through filter (A) and collect in the base of the unit at (F). This is shown in Figure 6.27. The inlet valve will remain open until maximum reservoir pressure is reached.

When the pressures above and below the diaphragm balance, the inlet valve will close, and exhaust valve (K) remains closed. This is shown in Figure 6.28.

When, due to normal usage of components or by natural wastage, the air pressure within the reservoir drops by 2 lbf/in^2 (14 kN/m^2, 0.1 bar), the pressure below the diaphragm in the drain valve will be greater than that above the diaphragm. Consequently, the

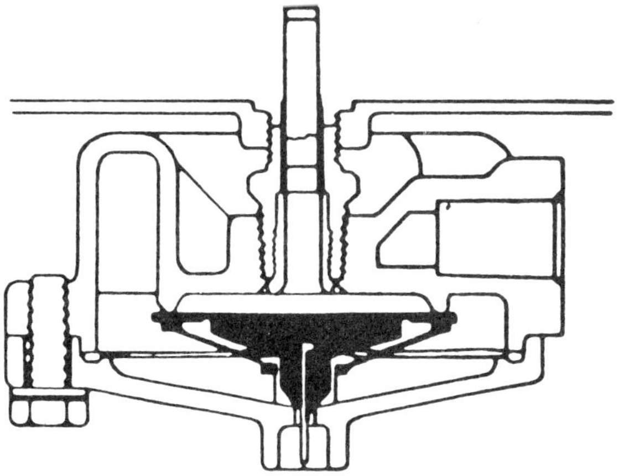

Figure 6.26 Automatic drain valve (DV2) – vehicle reservoir at atmospheric pressure

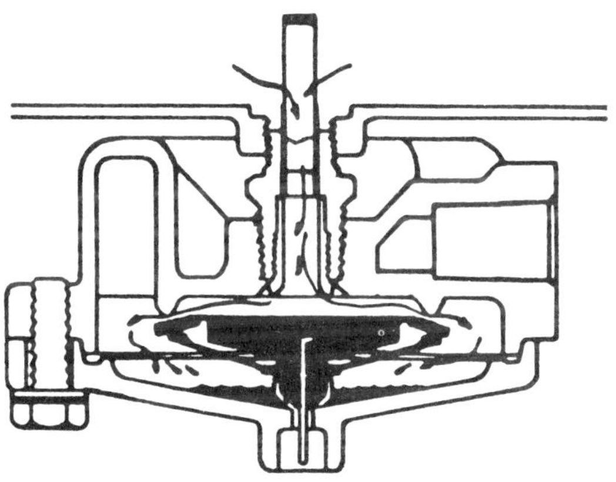

Figure 6.27 Automatic drain valve (DV2) – slight pressure rise in vehicle reservoir

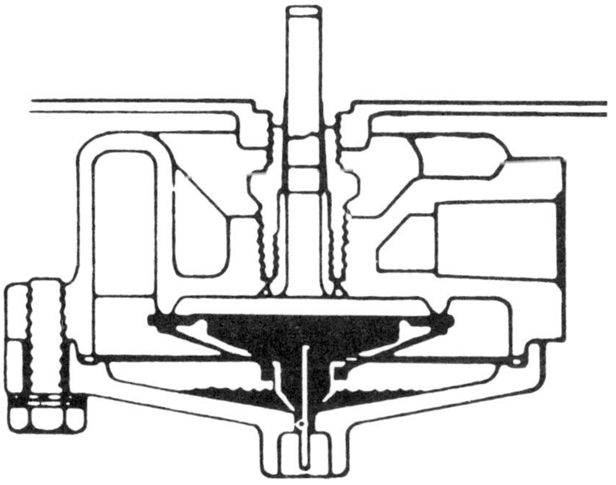

Figure 6.28 Automatic drain valve (DV2) – slight pressure drop in vehicle reservoir

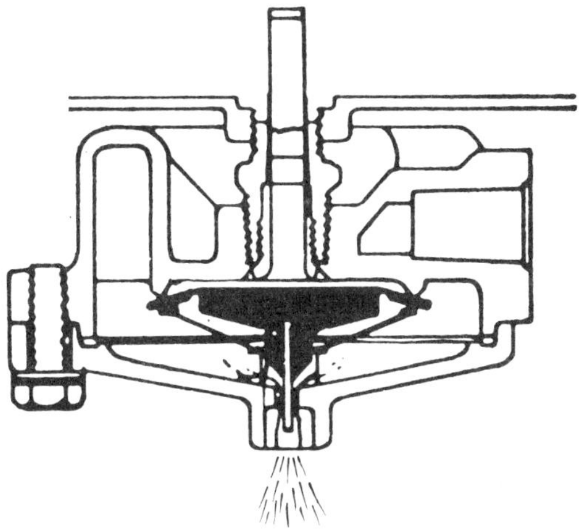

Figure 6.29 Automatic drain valve (DV2) – discharging moisture

pressure below, lifts the centre of the diaphragm and by so doing opens exhaust valve (K), expelling any condensate to atmosphere. This is shown in Figure 6.29. Pressure below the diaphragm will then fall and the exhaust valve will again close.

Manual draining is possible by depressing the centre wire at the base of the unit in a similar way to releasing air from a Schrader valve.

Automatic drain valve (KR)

This unit is fitted to a wet tank, that is installed between the compressor and the unloader valve.

As shown in Figure 6.30, port (1) is connected to a wet tank, and condensate from the wet tank flows down into (a) of chamber (A).

When the braking system unloader valve is in the charging condition, compressed air enters this automatic drain valve, via port (1), and acts upon piston (B), compressing return spring (C).

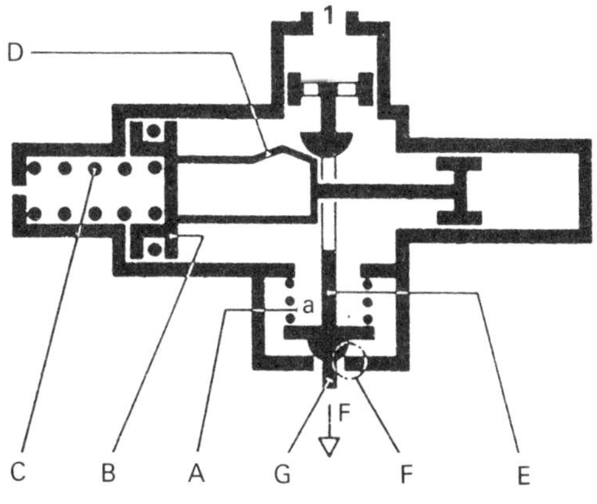

Figure 6.30 Automatic drain valve (KR)

When the braking system is fully charged, and the unloader valve cuts out, pressure at port (1) is reduced. Return spring (C) moves piston (B) to the right, causing cam (D) to lift valve member (E) and thereby opening exhaust outlet (F), so discharging the condensate. Manual draining can be achieved by pushing pin (G) upwards.

Electrically-operated automatic drain valve (HX)

Unlike most automatic drain valves, which depend for their operation on pressure changes within the air brake system for their discharge function, this valve is triggered electrically (see Figure 6.31).

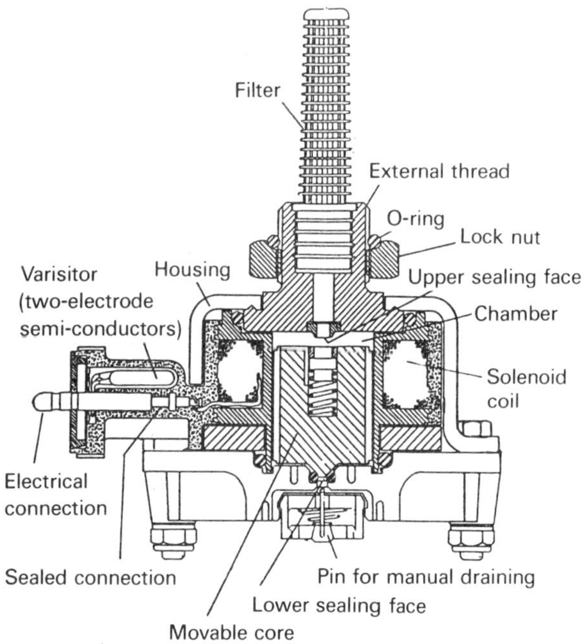

Figure 6.31 Electrically-operated automatic drain valve (HX)

Whenever the vehicle's electrical systems are switched on, the solenoid coil generates a magnetic field and causes the movable core to rise, sealing off the connection with the reservoir. Whenever the foot valve is applied, the electrical supply to the solenoid coil is broken. Consequently, the magnetic field collapses and the movable core is moved down by compressed air and gravity. Condensate in the reservoir flows into the chamber.

When the footbrake treadle is released, the electrical circuit to the solenoid coil is reinstated and hence the magnetic field again attracts the movable coil to its highest position. This unseats the lower sealing face which opens the exhaust port to discharge any liquid condensate to atmosphere.

This valve is produced with a heating element to prevent freezing in extremely cold conditions.

Automatic drain valve (WO)

This type of drain valve is *not* screwed directly into a reservoir, but it is mounted in the pipework of the system at a height, for port (2), not exceeding 300 mm above the reservoir drain boss (see Figure 6.32).

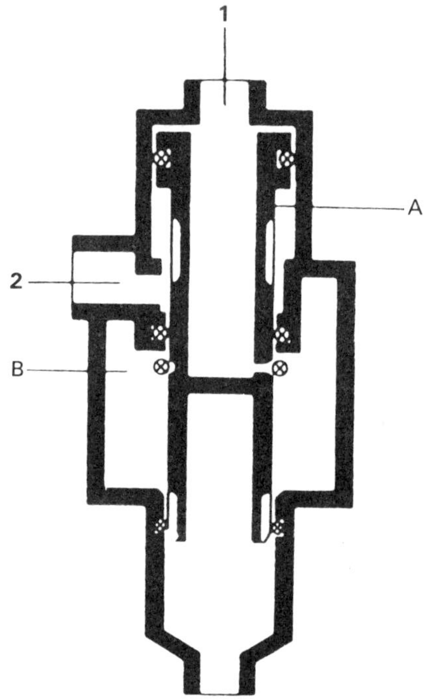

Figure 6.32 Automatic drain valve (WO)

Compressed air from an auxilliary port on the unloader valve is connected to port (1) of the automatic drain valve, and pressure at this point acts upon piston (A) moving it downwards, to its lowest possible position. Compressed air (and moisture) from the reservoir, enter port (2) and pass into chamber (B), via the reduced diameter section of piston (A). Moisture in the control line (i.e. the line from the unloader valve) entering port (1) can also enter chamber (B) via a small drilling in piston (A).

When the unloader valve cuts-out (at design pressure), the pressure at port (1) drops to nil. At this stage the reservoir pressure moves piston (A) upwards to its highest level, so that the reduced diameter of that piston, at its base, provides an exhaust route for the liquid in chamber (B), which is ejected to the atmosphere, by the trapped compressed air in that chamber.

Note: There is an O-ring covering the small drilling in piston (A), to prevent air or moisture flowing back through port (1), to the control line.

Safety valves (BW and WO)

A safety valve is always fitted into a compressed air braking system to prevent the build up of excessive pressures, which could occur if the governor or unloader mechanisms should fail. It can be fitted in a choice of locations, such as on the compressor cylinder head, on the unloader valve or on a reservoir.

The two examples shown in Figures 6.33 and 6.34 rely in the first case on a spring-loaded ball in a brass housing, and in the second case on a valve and diaphragm.

Under normal working conditions for the braking system, the safety valve does not come into operation. However, if the pressure beneath the valve becomes excessive, this will cause the ball to lift off its seat in the first case, or the diaphragm to lift and unseat the valve in the second case, and allow the pressurised air to exhaust to atmosphere. When normal pressure is restored, the regulating spring reseats the ball, sealing the exhaust port.

The regulator spring tension can be adjusted to screw action, but care should be taken to keep within

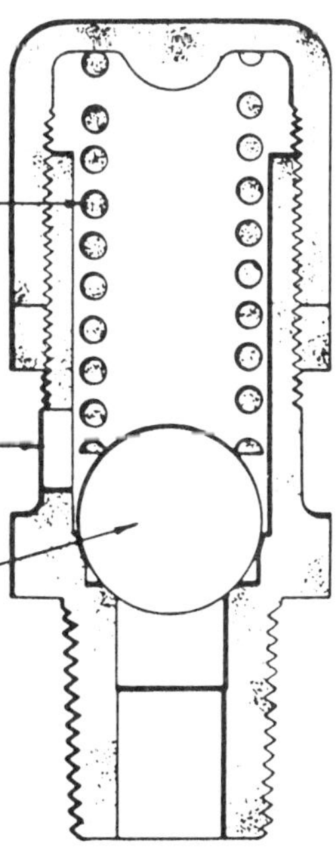

Figure 6.33 Safety valve (BW)

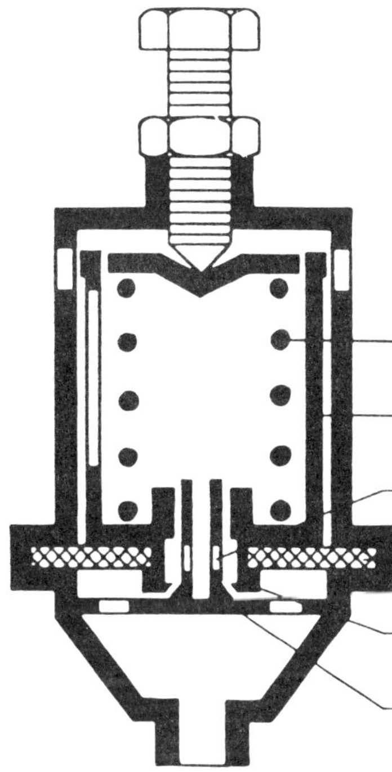

Figure 6.34 Diaphragm safety valve (WO)

the tolerances that are laid down for the particular system by the manufacturers.

It is usual for the system to be set to discharge to atmosphere between 130–160 lbf/in^2 (896–1103 kN/m^2, 9–11 bar).

If a safety valve is discharging, check the cab gauges. If these are showing an abnormally high pressure, adjust the unloader or governor valve, in accordance with the manufacturer's instructions. If the gauge pressure is not above normal, the safety valve is defective and should be serviced or replaced.

Whichever cause is to blame for such discharging, this defect should be rectified because continual losses lead to the overworking of the compressor, with the consequent reduction of its useful life.

Unloader valves

An unloader valve is fitted in the air brake system between the compressor and the reservoir, to maintain the air pressure in the reservoir within specified limits. When the maximum limit is reached, and this is usually set between 95–120 lbf/in^2 (655–827 kNm^2, 6.5–8.3 bar) the valve cuts out and exhausts the air being supplied by the compressor, to atmosphere. When the pressure in the reservoir drops to cut-in pressure, usually set by the manufacturer at 15 lbf/in^2 (103 kN/m^2, 1 bar) below the cut-out pressure, the feed from the compressor to the reservoir is reinstated. The differential between the cut-out and cut-in pressures, set in manufacture, cannot be altered, but the cut-out pressure and, consequently, the cut-in pressure can be adjusted in service by the use of the screwed bush that acts upon the control spring.

Unloader valve (piston-type) (BW)

As shown in Figure 6.35, air from the compressor enters the unloader valve through chamber (A) and passes on to the reservoir, via a check valve within this component. Reservoir pressure, beyond this check valve, is also fed through drilling (B) to act upon the pilot valve piston at (C). This pressure also reaches port (E), which can be used for fitting a safety valve, low-pressure indicator, pressure gauge or can just be blanked off.

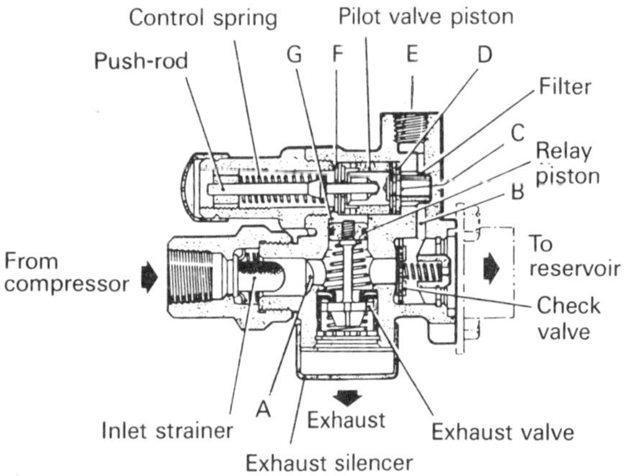

Figure 6.35 Unloader valve (piston type) (BW)

The exhaust valve and the relay piston are held up by the springs below them, and the exhaust valve is firmly seated by air pressure, when this is present in chamber (A), which acts upon the large area underneath the relay piston. Chamber (G), above the relay piston, is vented to atmosphere via the pilot valve seat (F), push-rod and the control spring housing.

As the reservoir pressure rises, the same pressure is also acting upon the pilot valve piston at (C), and the setting of the control spring determines the amount of pressure required to move this piston towards the seat at (F). The moment the pilot valve piston moves off its seat at (D), the larger area of the pilot valve is exposed to the air pressure which causes the valve to snap onto its seat at (F), thus sealing off chamber (G) from atmosphere. Reservoir pressure is now felt on top of the relay piston, momentarily equalising the pressure felt underneath. All the effective pressure in chamber (A) is now acting upon the exhaust valve, which will commence to open. The compressor will only pump to atmosphere through chamber (A) and exhaust, as long as the reservoir pressure remains high enough to keep the pilot valve piston against seat (F).

When the reservoir pressure acting on the pilot valve piston drops below the control spring setting, the spring pressure will move the piston back towards seat (D). Immediately the piston leaves seat (F), the pressure in chamber (G) can escape to atmosphere. The springs under the exhaust valve and relay piston, return them to the closed position. At this stage the compressor can again recharge the reservoir.

Unloader valve (WO)

As shown in Figure 6.36, air from the compressor enters port (1) and then passes through filter (14) and chamber (B), before leaving port (21) via non-return valve (9), en route to the storage reservoirs.

At the same time, a tell-tale sample of this air pressure enters chamber (A) and acts upwards beneath diaphragm (8), which reacts against piston (7) and return spring (6).

When air pressure acting upwards beneath diaphragm (8), exerts a greater force than the downward force exerted by spring (6), the diaphragm, piston and inlet valve (19), all move up. This action closes exhaust (20) and opens inlet (19), allowing com-

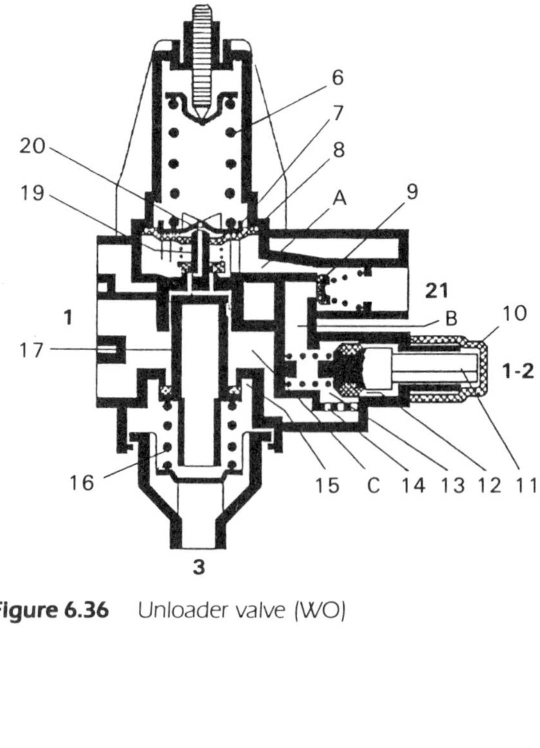

Figure 6.36 Unloader valve (WO)

pressed air from chamber (A) to act down upon piston (17), against the resistance of spring (16), consequently opening valve (15). This allows incoming air from the compressor and air from chamber (C), to escape to atmosphere via port (3). Non-return valve (9) closes owing to a drop in air pressure.

This particular unloader valve can be fitted with a tyre inflator which operates as follows. When dust cap (10) is removed, and the union nut of the hose is screwed onto the threaded port, valve (11) is pushed off its seat (12), sealing the inner end of the valve against seat (13). Air from the compressor enters port (1) and chamber (A), then passes through filter (14), and port (1-2) to the inflator hose.

Unloader valve (diaphragm-type) (CD)

This valve has three working barrels, two parallel to each other and one at 90° to the other two barrels.

As shown in Figures 6.37 and 6.38, compressed air from the compressor, travels through the left-hand barrel, passing through a non-return valve before continuing towards the reservoir. Simultaneously, compressed air passes through a drilling in the body, on the reservoir side of the non-return valve, to reach an area beneath the governor diaphragm (in the barrel which is at 90° to the other two). At this stage the unloader plunger in the right-hand barrel of the parallel pair is being held onto its seat by spring pressure.

As the pressure in the reservoir builds up, this pressure is also present in the drilling and beneath the governor diaphragm, causing the diaphragm to lift against the resistance of the pressure-setting spring. The air pressure is also being communicated to the

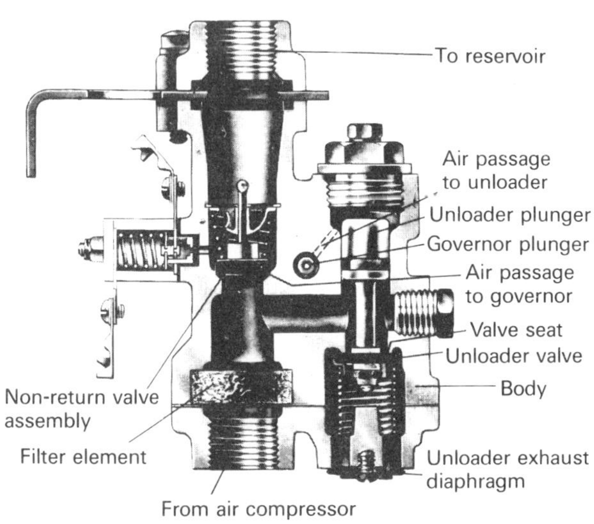

Figure 6.37 Unloader valve (diaphragm type) (CD)

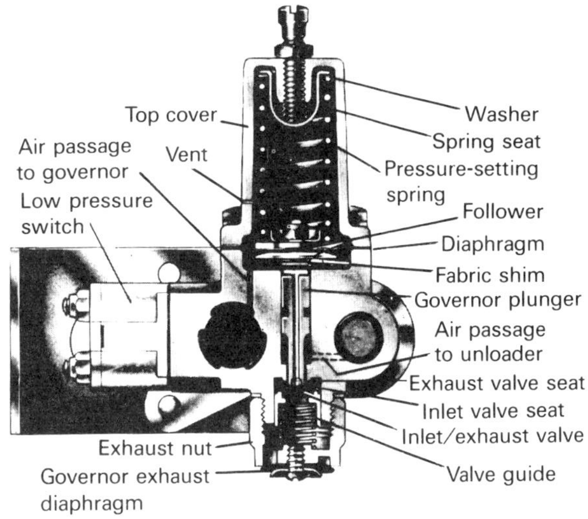

Figure 6.38 Unloader valve (diaphragm type)

hollow plunger beneath the diaphragm, the base of which is sealed by the spring-loaded inlet/exhaust valve. When the pressure in the reservoir reaches a predetermined figure, according to the adjustment of the pressure-setting spring, the spring-loaded inlet/exhaust valve closes the exhaust passage in the body and the inlet valve opens. Thus compressed air passes down through the governor plunger, through a drilling in the body, to reach the area in the bore above the unloading plunger, in the right-hand barrel in the parallel pair.

At this stage, air pressure is also acting upwards on the lower section of the governor plunger, helping to move the diaphragm upwards and also fully opening the inlet valve. When the inlet valve opens, the air pressure, passing through a drilling in the body, moves the unloader plunger in the right-hand bore, downwards against spring pressure, unseating the unloader valve and allowing air coming from the compressor to exhaust to atmosphere.

When, as a result of the driver's usage of components or by natural wastage, the pressure in the reservoir drops to cut-in pressure, the spring pressure above the governor diaphragm will be greater than the reduced air pressure below. The governor diaphragm and plunger will be moved downwards, closing the inlet valve and opening the exhaust valve, so that air above the unloader plunger, in the right-hand barrel, can escape to atmosphere via the drilling through the body and the open exhaust in the base of the bore containing the governor plunger. Spring pressure will reseat the unloader valve, closing its exhaust. The reservoir can be recharged to the preset pressure limit, when the unloading will again commence.

Unloader valve (WO)

As shown in Figure 6.39, compressed air from the compressor enters port (1) and passes through filter (14). It then passes through chamber (e) and non-return valve (9) before leaving port (21) en route to the reservoirs. Whilst this process is taking place, some of the compressed air is also entering chamber (g) and acting upwards under piston (19) and diaphragm (20), which are being restrained by pressure spring (6). (There is an adjusting screw (5) and lock nut, which enables the cut-out pressure for the system to be varied.)

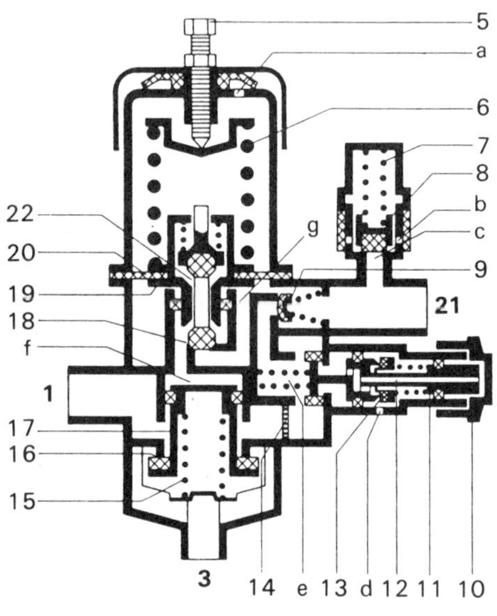

Figure 6.39 Unloader valve (WO)

When the cut-out pressure is reached the dumb-bell shaped inlet/exhaust arrangement is seated at exhaust port (22) and unseated at inlet port (18), thus enabling air in chamber (g) to pass through to chamber (f). It then acts downwards upon piston (17), against the force of spring (15), opening exhaust valve (16) and allowing air to escape to atmosphere through exhaust port (3).

This situation will continue until, due to the driver's use of brake valves and/or by natural wastage, the pressure in the reservoirs drops to below design pressure (cut-in pressure). As a result, piston (19) and diaphragm (20) are returned to the horizontal position, by spring pressure, which again exceeds the air pressure that is acting upwards. The inlet port (18) is closed and the exhaust port (22) opens, and air in chamber (f) exhausts to atmosphere via drilling (a). Piston (17) is forced upwards by return spring (15), thus closing exhaust (16), and consequently incoming air from the compressor passes to the storage reservoirs, via filter (14) and non-return valve (9), until the system reaches cut-out pressure, when the whole procedure is repeated.

Safety valve

As shown in Figure 6.39, air passing through drilling (c), acts against valve body (8). When the preset value of this valve is reached, the body moves, against return spring (7), to discharge air to atmosphere via vent (b).

Tyre inflator

As shown in Figure 6.39, to use the tyre inflater, wing nut (10) is unscrewed. As the union nut of the tyre inflater hose is screwed into place, the tappet (11) is moved to the left, thereby closing the route between chamber (e) and reservoir port (21). In this condition compressed air delivered by the compressor passes from chamber (e), through drilling (12) to the inflation hose and tyre. Should the pressure in the tyre being inflated by the hose exceed 145 lbf/in^2 (1000 kN/m^2, 10 bar), pressure will unseat relief valve (13) and escape to atmosphere via drilling (d).

Note: No inflation can occur if the vehicle's reservoirs are already fully charged, i.e. the unloader valve is discharging to atmosphere. In such circumstances the reservoir pressure must be slightly reduced, either by use of the foot/handbrake control or by limited use of reservoir drain valves.

When connecting the pressure hose to the Schrader valve of the tyre, if pressure in the hose exceeds approximately 145 lbf/in^2 (999 kN/m^2, 10 bar), the safety valve (8) would allow the excess pressure to escape to atmosphere.

Unloader valve (KB)

Operation of an unloader valve

As shown in Figure 6.40, air supplied by the compressor enters port (1) and passes through a filter and non-return valve, prior to leaving by port (21), en route to the storage reservoirs.

Reservoir pressure also passes through a drilling and acts beneath the control piston, which is restrained from above by a spring. As the air pressure increases the control piston rises, until the sealing ring passes the horizontal drilling in the valve tappet. This allows air to

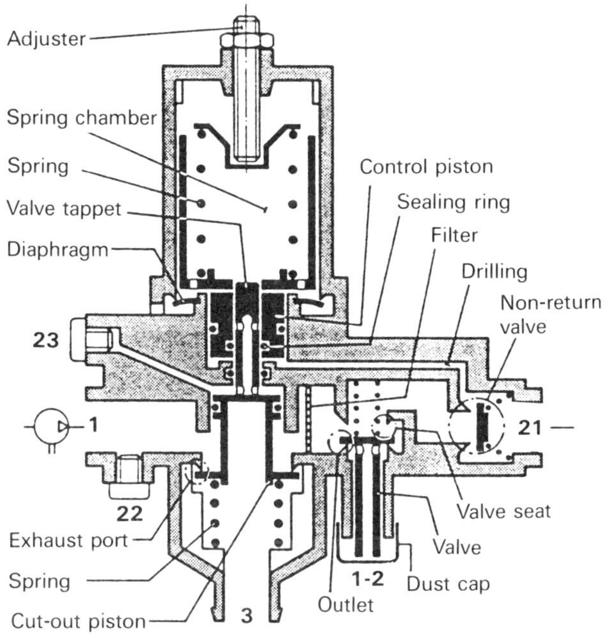

Figure 6.40 Unloader valve (KB)

pass through the drilling and act down upon the cut-out piston, causing it to move off its seat and allow the incoming air from the compressor to escape to atmosphere via the exhaust port.

When the air pressure in the storage reservoirs is reduced, the force of the spring overcomes the air pressure acting beneath the control piston and causes the sealing ring to be moved away from the cross-drillings in the valve tappet. The pressure acting above the cut-out piston is then exhausted to atmosphere, via the valve tappet, spring chamber and diaphragm. Consequently, the cut-out piston reseats and seals off the exhaust port.

Compressed air from the compressor is then, once again, supplied to the storage reservoirs.

Tyre inflator

As shown in Figure 6.40, to use the tyre inflator, it is necessary to remove the dust cap, from port (1-2), and then screw the inflator hose onto the external thread of port (1-2). This causes the spring-loaded valve to come into contact with the valve seat, thereby preventing the reservoir being discharged.

Compressed air from the compressor, enters the outlet port and the hollow stem of the valve, to reach the tyre inflation hose. Tyre inflation can only be achieved with the engine running, i.e. with the compressor running. It is also necessary for the unloader valve to be in the charging condition. Where the reservoirs are already fully charged, and consequently the unloader valve has cut-out, it is necessary to reduce the reservoir pressure. This will cause the unloader valve to cut-in and allow the system to commence recharging.

Safety

To minimise the risks when inflating tyres, the action of the unloader valve prevents the use of excessive air pressure. As shown in Figure 6.40, when air pressure within the unloader valve rises to the maximum (design) pressure, the cut-out piston moves down, overcoming the resistance of the spring and allowing air to escape to atmosphere via the exhaust port.

Charging the storage reservoirs from an external source

As shown in Figure 6.40, the tyre inflation port (1-2) can be used as a 'first aid' point, to feed compressed air from another vehicle or a workshop air-line, to the storage reservoirs of this vehicle. In such cases, the spring-loaded valve must not be forced against the valve seat. By adopting this action, a route is left available, via the outlet port, to enable the incoming air to reach the storage reservoirs of the vehicle.

Governor valve (diaphragm-type) (CD)

A governor valve is fitted in the storage section of the system, to maintain a working pressure, allowing the compressor to run light by discharging to atmosphere when that pressure has been achieved. The actual discharging to atmosphere is caused by an unloader mechanism situated on the compressor which is activated by signal pressure from the governor valve. The unloader mechanism, which should not be confused with an unloader valve, consists of a plunger operated by compressed air supplied by the governor, which is forced down to hold the inlet valve open, preventing compression taking place.

As shown in Figure 6.41, the governor valve receives compressed air from the reservoir at the inlet port. This air acts upwards on the underside of the diaphragm and downwards through the centre of the hollow plunger, which is sealed at the lower end by the spring-loaded inlet/exhaust valve. As pressure increases, the diaphragm assembly lifts causing the plunger below to rise, so that the exhaust valve seats and the inlet valve opens, allowing air to pass out of the port connected to the unloader mechanism on the compressor. As the plunger in the governor valve lifts from the inlet/exhaust valve, air also acts upwards on the lower area

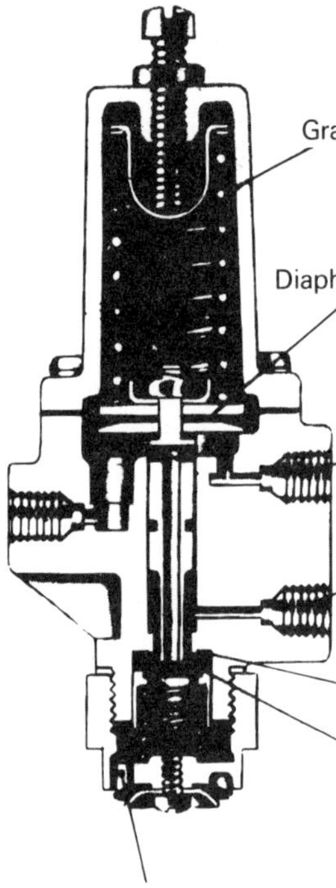

Figure 6.41 Governor valve (diaphragm-type) (CD)

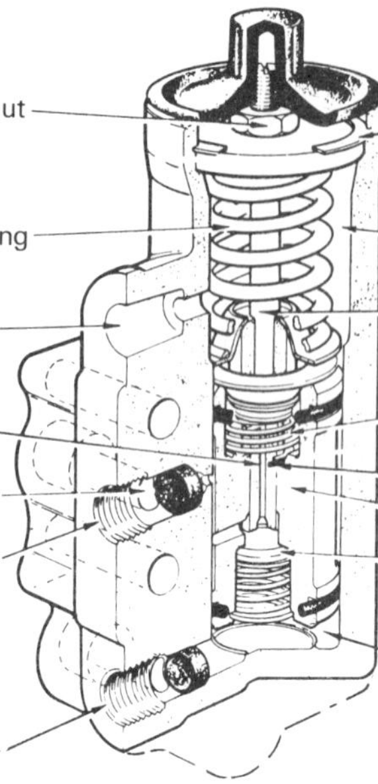

Figure 6.42 Governor valve (piston-type) (BW)

of the plunger, increasing the diaphragm area and fully opening the inlet valve on the compressor.

As the pressure in the reservoir decreases to the cut-in point of the governor valve, the diaphragm and plunger are forced back down, closing the inlet through the plunger and opening the exhaust valve. This allows the pressure to escape from the line to the unloader mechanism and the inlet valve on the compressor reseats. The compressor can then compress air until the cut-out point is reached, when the whole procedure will be repeated.

There is a connection on a governor valve for a pressure gauge or a low-pressure indicator.

Governor valve (piston-type) (D2-type) (BW)

This unit consists of a spring-loaded piston and a disc valve, within a cylindrical casting.

As shown in Figure 6.42, the bottom port connects chamber (A) to the reservoir, thereby ensuring that this port is always at the same pressure as the reservoir. Chambers (B) and (C) are connected to atmosphere via exhaust.

When the reservoir is empty or contains low pressure, the components within the governor are in the positions shown in Figure 6.42. As the pressure is built up the piston is pushed upwards, thus bringing the disc valve into contact with the spring-loaded exhaust system. As pressure continues to increase, the exhaust stem pushes the disc valve off its seat, allowing the compressed air from the reservoir to enter chamber (D) and pass on to the unloader port. The compressed air is then routed, by pipeline, to the unloader mechanism in the cylinder head of the compressor, which allows the compressor to run light whenever the design pressure has been achieved in the reservoir.

When the air pressure in the reservoir falls, the piston in the governor valve is moved downwards by the force of the control spring, thus allowing the disc valve to return to its seat on the piston, and so isolating the unloader mechanism from the reservoir pressure.

The exhaust stem is kept in contact with the adjusting screw by spring pressure and this arrangement allows the pressurised air in ports and passages, leading to the unloader mechanism, to exhaust to atmosphere via the hollow exhaust stem, to holes in lower spring seats, and thence to the exhaust port.

Non-return valve (BW)

This type of valve, often described as a *check valve*, is fitted to the inlet port of all reservoirs on older vehicles. When a sensing tank is fitted, in series, between the unloader valve and the main reservoir, a check valve is normally fitted to the reservoir only. Check valves are not fitted to tanks on trailers, because air must be allowed to flow in both directions in order that the relay emergency valve may operate, in an emergency situation, to apply the trailer brakes.

As shown in Figure 6.43, a non-return valve consists of a body, with a screwed cap, containing a rubber valve (either dome-shaped or disc), a spring and guide.

The pressure of air supplied by the compressor forces the valve off its seat, against spring pressure, allowing entry to the reservoir. Reverse flow of air is prevented by spring action upon the valve, aided by the pressure in the reservoir.

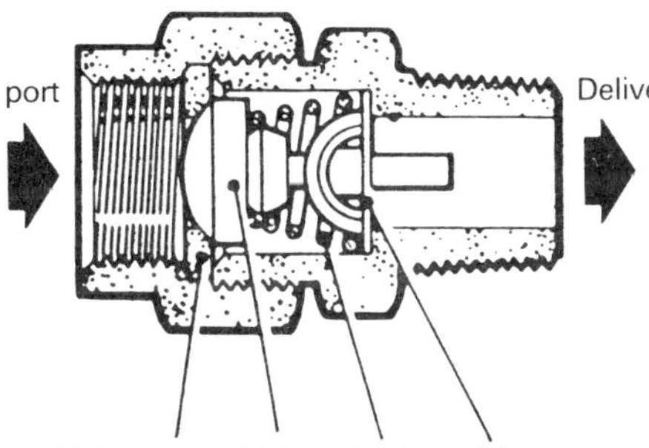

Figure 6.43 Non-return valve (BW)

With the advent of EEC regulations, the non-return valve, in its present form is being replaced with a *single circuit* or *system-protection valve*, which will protect the remainder of the system against losses resulting from the failure of one part of the system.

System-protection valves (CD)

The system-protection valve (CD) and a similar valve, the circuit-protection valve (BW), have replaced the conventional non-return valve (check valve) that was normally fitted to the inlet port of reservoirs. These later styles of valve still contain a non-return valve, but they also permit reservoirs that have not failed, and their systems or circuits, to operate. This is an EEC requirement, as on the Continent vehicles are allowed to be driven back to base if one braking system has failed. At present that allowance is still a contravention of UK regulations.

These valves are produced as single-, triple- or quadruple-element types. The multiple types allow a reduction in pipework and also incorporate non-return valves.

However, conventional non-return valves are still used in other parts of air brake systems to prevent a reverse flow of air.

Single-element system-protection valve

As shown in Figure 6.44, compressed air being delivered by the compressor presses upon the face of the disc valve, lifting it when the pressure overcomes the setting spring. As the valve starts to lift, the air pressure is then able to act on the larger area of the valve then exposed, opening the valve further. As the pressure in the reservoir, beyond the valve, builds up, it acts upon the piston, allowing the valve to open fully. When the unloader valve, or governor, cuts off the supply of compressed air, the small spring and the air pressure in the reservoir, cause the non-return valve to reseat, preventing back flow of air. The reservoir pressure still holds the piston in the raised position.

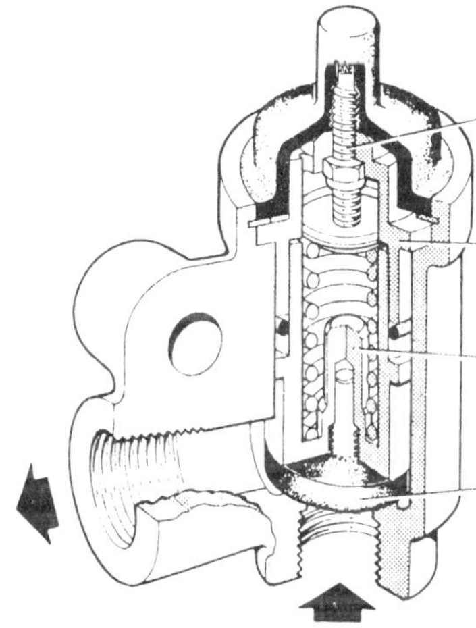

Figure 6.44 System or circuit protection valve (single-element)

When, as a result of use of components or by natural wastage, the reservoir pressure drops, the unloader arrangement allows charging to restart. The non-return valve opens and the reservoir is again charged. If the air pressure in one system drops by 30 lbf/in^2 (207 kN/m^2, 2 bar), the air pressure acting upon the valve and piston will not be sufficient to overcome the pressure of the setting spring, so the valve will close. Although the remaining pressure in the reservoir will be acting upwards upon the piston, the area exposed to this air will be reduced by the valve being in the closed condition and an increase of 25 lbf/in^2 (172 kN/m^2, 1.7 bar) may be necessary to open the valve.

If a fault causes a total loss of air in the reservoir of one system, the setting spring in the protection valve concerned closes the piston against the non-return valve. Braking can be achieved with the other circuits, as the pressure will be maintained at 92 lbf/in^2 (634 kN/m^2, 6 bar) and the valve on the faulty circuit will not open until the predetermined pressures have been achieved. The other sound circuits will also be able to function, as the residual pressure in their reservoirs will be keeping the pistons raised. Thus only a small pressure will be necessary to allow charging to commence.

One version of this valve has a feedback capability, so that the reservoirs to which these valves are fitted can equalise their pressure.

Multi-circuit protection valve (BW)

Manufacturers differ in their name for this valve and it is often described as a *multi-system protection valve*. This valve is a compact unit that is used as an alternative to four separate circuit protection valves.

There are two types of protection valve elements, namely feedback and non-feedback. A failure in a circuit downstream of either type of valve, automatically causes the element to isolate that circuit from the supply and to keep the air pressure high enough, in the remaining circuits, to achieve the minimum legal braking efficiency.

With the feedback type, the element remains open above closing pressure. With the vehicle in normal running condition, i.e. with no faults, and the reservoirs fully charged, the valves allow reservoir pressure to equal supply pressure. Thus, the feedback valve allows the compressor governor valve to sense the pressure in all of the reservoirs, without the need for a sensing tank.

Non-feedback valves are fitted in the supply lines to other reservoirs, and these act as reservoir check (non-return) valves, which remain closed, unless the respective reservoirs are being charged with compressed air.

As shown in Figure 6.45, two single-check valves are fitted in the airways between circuits (1) or (2) and valves (3) and (4). If a circuit and/or reservoir should fail, downstream of either valve (1) or (2), the appropriate single-check valve will act to prevent the supply of air to valves (3) and (4), bypassing the affected valve element and failed circuit. Valve elements (3) and (4) will continue to receive compressed air, via the unaffected valve element, up to the opening pressure of the affected valve element.

This valve has five ports, all of which are connected to internal air passages that connect two pairs of valve seat chambers, (B) and (C), and check-valve chamber (D). Four protection valve elements are mounted, in pairs, vertically opposed.

The valve shown contains one feedback valve (F) and one non-feedback valve (G) in each pair of valves. Each valve has a rubber seal which contacts the valve

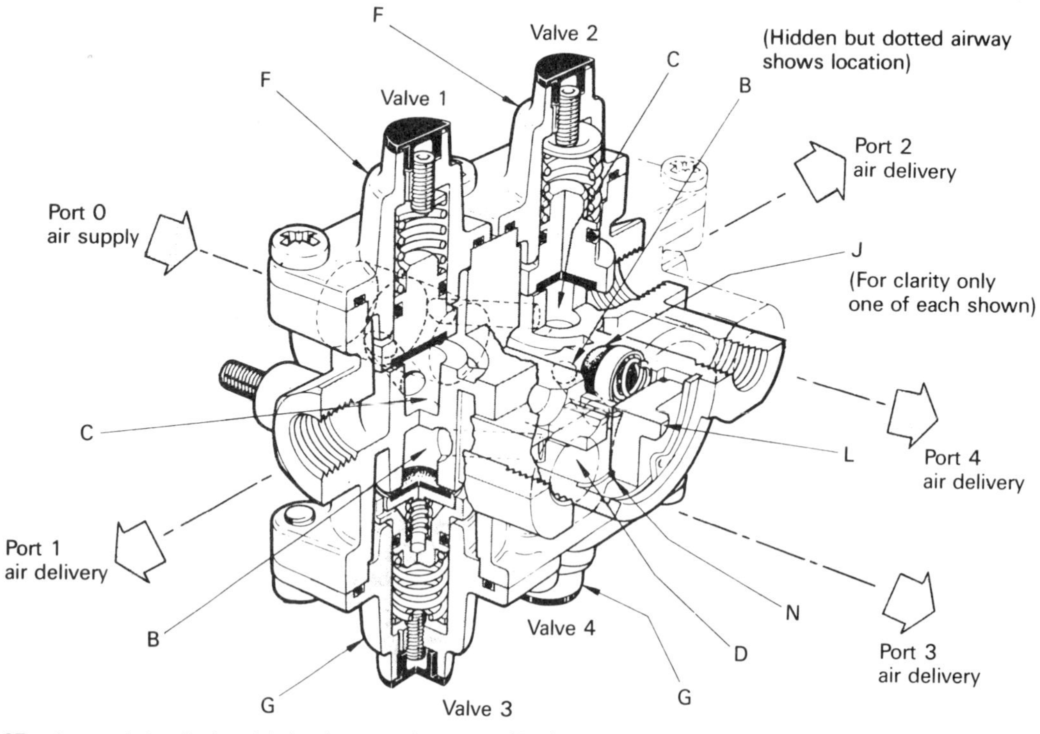

Figure 6.45 Internal detail of multi-circuit protection valve (BW)

seat of chambers (B) and (C). The two check valves (J) are retained by plate (L) and sealed by O-ring (N).

It can be seen that port (0) provides an entry point for compressed air from the compressor. The internal airways divide, enabling the incoming air to enter valve seat chambers (C) at valves (1) and (2). The airways then again divide, one feeding delivery port (1) or (2), whilst the other airway is routed to chamber (D) via check valve (J). Two further airways link this chamber with valve seat chamber (B) at valves (3) and (4). Beyond these valves, airways supply compressed air to delivery ports (3) and (4).

As shown in Figure 6.46A, a feedback-type valve element lacks check valve (S) and spring (U). Valve element (R) is bonded onto the end of piston (T). Both types are fitted with screw (W) to enable the piston spring tension to be adjusted. Breather cap (X) and airway (Y) allow the piston chamber to be vented to atmosphere.

As shown in Figure 6.46B, rubber disc valve (R) is bonded to check valve (S), which is free to move within the end of piston (T). Check valve spring (U) acts against check valve (S) and sealing element (R), when piston (T) rises against piston spring (V). The crown of piston (T) is slotted to allow compressed air to act upon the complete area of the piston end.

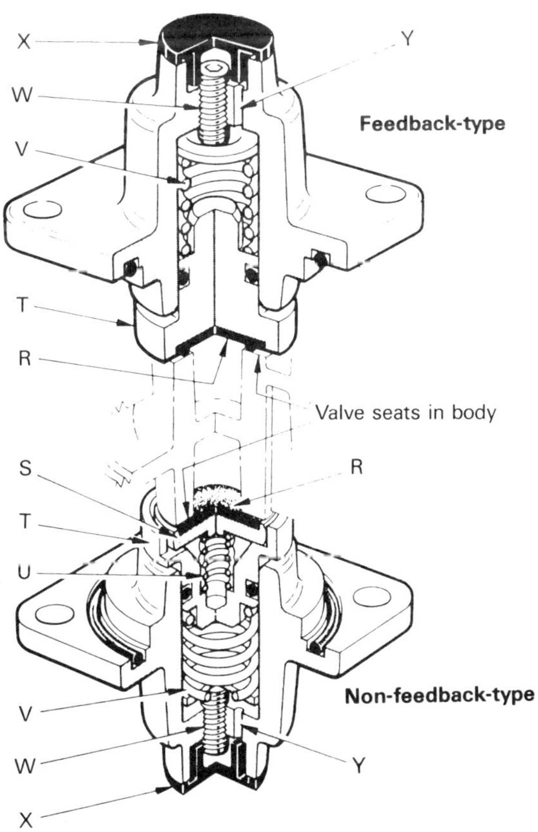

Figure 6.46A Feedback-type valve elements

Figure 6.46B Non-feedback-type valve elements

Operation (see Figures 6.45 and 6.46A and 6.46B)

As shown in Figure 6.45, compressed air entering port (0) passes to valve seat chambers (C), under the feedback valve element (F) (valves (1) and (2)).

As shown in Figures 6.46A and 6.46B, when air pressure reaches the pressure settings, for example approximately 84 lbf/in^2 (580 kN/m^2, 5.8 bar), piston spring (V) and feedback piston (T) lift, and compressed air flows through to the service brake reservoir, via delivery ports (1) and (2). It also flows past check valves (J) into check-valve chamber (D). Incoming air then acts upon the complete face area of the feedback pistons, keeping them in the raised position to allow service reservoir pressures to build up to the supply pressure.

A failure in a service brake circuit, causing the reservoir pressure to fall, will cause the relevant feedback valve piston to move down until, at approximately 58 lbf/in^2 (400 kN/m^2, 4 bar), depending upon the pressure setting, it fully closes the feedback valve air inlet. Check valve (J), in air route from chamber (D), will also close, to prevent loss of supply air from the unaffected feedback valve. Supply air pressure will then rise until the affected valve begins to open. The sound reservoirs/circuits continue to receive compressed air supply, up to this opening pressure of the affected valve.

Incoming supply air flows through chamber (D) to the valve seat chambers beneath the non-feedback valve element (G) (valves (3) and (4)). When the supply air reaches the setting pressure of non-feedback piston springs (V), for example 84 lbf/in^2 (580 kN/m^2, 5.8 bar), non-feedback check valves and pistons (S) and (T) lift simultaneously, allowing air to pass to their reservoirs via delivery ports (3) and (4). Air then acts upon the full face area of the non-feedback pistons, causing them to rise against their return springs. When the air supply ceases, for example, when the governor valve causes the compressor to run light, check valves (S) close the non-feedback valve air inlets, acting as reservoir check valves, preventing cross-bleeding from one non-feedback valve to the other via chamber (D). Check valve (J) prevents chamber (D) from allowing air to flow back to circuits (1) and (2) if either of these circuits suffer a pressure loss.

In the event of pressure failure in either reservoir supplied by ports (3) and (4), the affected non-feedback piston (T) will travel down, until at approximately 58 lbf/in^2 (400 kN/m^2, 4 bar), depending upon the setting, it closes off the non-feedback valve air inlet. Supply air pressure will then rise to the opening pressure of the affected valve and remaining reservoirs/circuits will receive pressure, up to that amount, controlled by the adjusting screw (W).

Reservoirs

These are storage tanks for compressed air, held in readiness for when it is needed. In addition, they can provide a reserve in case of compressor failure so that the brakes can still be applied. They should be drained regularly to get rid of oil and water that may have entered the system.

The testing of all reservoirs, by manufacturers, is very rigorous. Although safety valves on vehicles are set to unload at about 160 lbf/in^2 (1103 kN/m^2, 11 bar), reservoirs are tested on the production line to withstand 250 lbf/in^2 (1723 kN/m^2, 17 bar). Periodically, reservoirs are taken from the line and tested to destruction. The minimum requirements for 8 in (203 mm) diameter tanks is to withstand a pressure of 1000 lbf/in^2 (6894 kN/m^2, 69 bar) and for 10 in (254 mm) diameter tanks to withstand 850 lbf/in^2 (5860 kN/m^2, 59 bar). Under this hydraulic testing these tanks usually withstand between 1200 and 1400 lbf/in^2 (8272–9652 kN/m^2, 83–97 bar) before failing.

Some modern braking systems are designed to store compressed air at a pressure of 250 lbf/in^2 (1723 kN/m^2, 17 bar) and naturally, suitable safety checks are made during the production of these reservoirs. The purpose of such high storage pressures is to reduce the size and number of reservoirs required on a vehicle. However, the pressure is reduced to normal usage pressures when delivered to control valves and actuators, etc.

The working life of a reservoir is calculated as being about ten years, providing the operator applies sensible maintenance. Dents not only reduce the capacity of a reservoir, but also provide a potential weakening of the tank. It should be noted that some modern reservoirs are constructed of aluminium. If any reservoirs are in a position that require fitters or drivers to walk or climb onto them, at the rear of the cab of an articulated tractor, some form of catwalk should be permanently in position.

Apart from periodically checking for leaks, for security of fastening to the chassis and for corrosion, reservoirs will cause little trouble.

Wet tanks

These very small tanks are often fitted into systems. The object is to collect the moisture brought in from the atmosphere, after it leaves the compressor and before it reaches the air reservoirs, and to discharge it back to atmosphere, usually through an automatic drain valve. An alternative to this arrangement is the use of an air drier or condenser.

Draining reservoirs

Although there is a maintenace section at the rear of this publication, as the draining of air tanks on vehicles is an ever-present problem, it is highlighted at this point. The job involved is not the problem as it is very simple; it is trying to get both drivers and fitters to accept that it is their responsibility to do this task. A couple of minutes work daily (or even weekly) can lengthen the life of the braking system, and is indeed a very important safety procedure.

Water washes off lubricant, accelerates corrosion and in cold weather can freeze-up components. A seriously neglected system can accumulate so much water that the actual air capacity of a reservoir is reduced (water does not compress).

Figure 6.47 shows the basic manual drain valve and Figure 6.48 shows a drain plug. Drain plugs *are* still fitted to some vehicles.

Warning: As the plug is being undone, high air pressure is still present in this reservoir and consequently as the last thread is reached, the plug can be ejected violently by the air pressure within the reservoir.

Figure 6.47 Drain cock

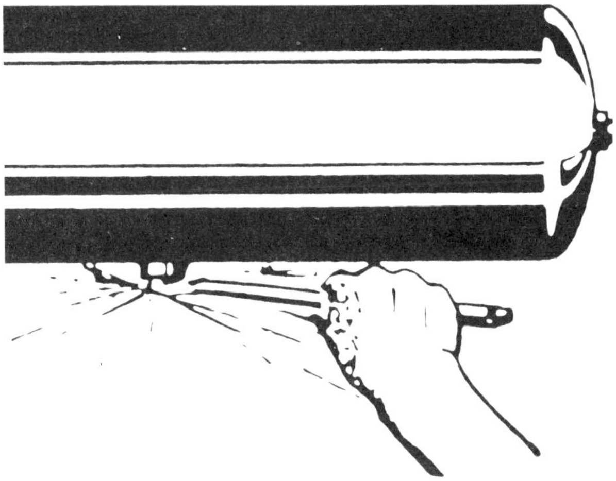

Figure 6.48 Drain plug

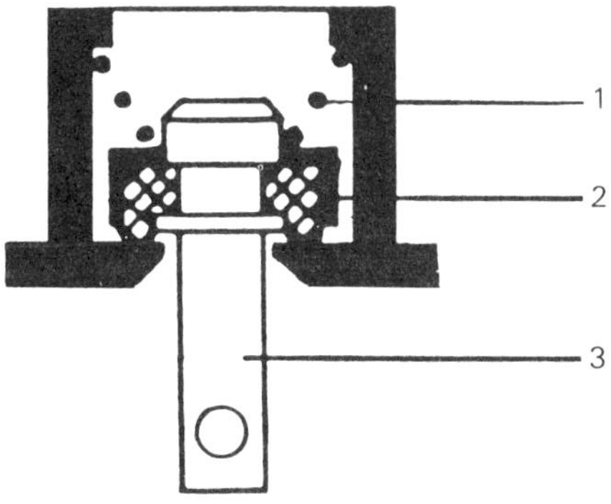

Figure 6.50 Tilt drain valve (WO)

Tilt drain valve (WO)

This is a manually-operated drain valve fitted to a reservoir. It has been devised to aid (and encourage) the regular draining of reservoirs. It is a substitute for the orthodox drain cock, especially where the reservoir concerned is inaccessible. Figure 6.49 shows draining with a tilt valve.

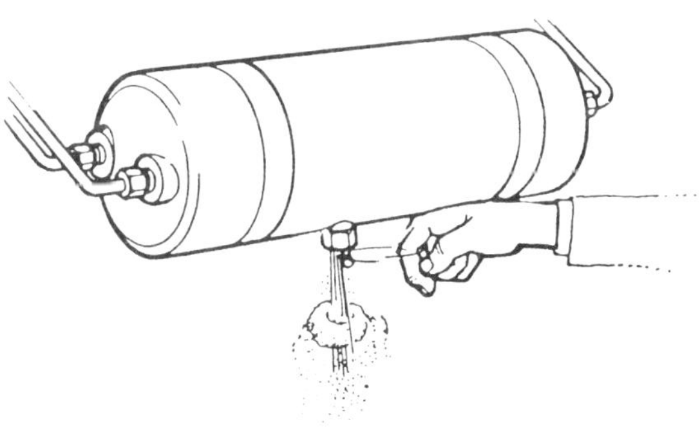

Figure 6.49 Draining with a tilt valve

As shown in Figure 6.50, the operating principle of this unit is the use of a horizontal, spring-loaded tilt valve (2), that when at rest, seals off the exhaust port. An actuating pin (3), vertically-mounted, passes through, and is sealed to the tilt valve. When the actuating pin is moved away from its vertical position, the tilt valve follows, thus unseating its lower face from the exhaust port, and allowing pressure to escape from the reservoir (1) above this unit. This escape will continue for whatever length of time the pin and valve are tilted.

There is a drilling in the actuating pin to receive a wire which when fitted to the vehicle or trailer can be operated, via a pull loop, by the fitter or driver standing alongside that vehicle or trailer.

Air pressure gauges (WO)

The Bourdon tube principle is used in these instruments. As shown in Figure 6.51, an oval section tube, that has been fashioned into the shape of a question mark, is fastened at its vertical stem to the body of the instrument, whereas the curved section is not fastened. Consequently, when compressed air is fed into the vertical stem, the curved section tends to straighten. A lever and quadrant are attached to the free end of the curved section, with the teeth of the quadrant in mesh with the teeth of a gear attached to a pointer.

Convenience sometimes causes truck manufacturers to fit dual gauges, which merely use two separate input ports and two Bourdon tubes, with two separate needles operating on two separate (identical) scales on *one* gauge.

Low-pressure switch (BW)

This is a low-pressure warning device which is a safety factor and a legal requirement on most compressed air brake systems. As shown in Figure 6.52, it consists of a rubber cover, nylon cover and an aluminium body. There are two contact plates with terminals, a contact spring and contact disc, nylon piston and rubber diaphragm. Projections on the nylon piston pass through holes in the lower contact plate, keeping the contact disc away whilst the system is in operation. Air pressure entering this unit, acts upon the diaphragm below the piston. As the piston is moved upwards it breaks the contact between the electrical disc and plate, so breaking the circuit and stopping the buzzer. Conversely, when the pressure in the reservoir drops,

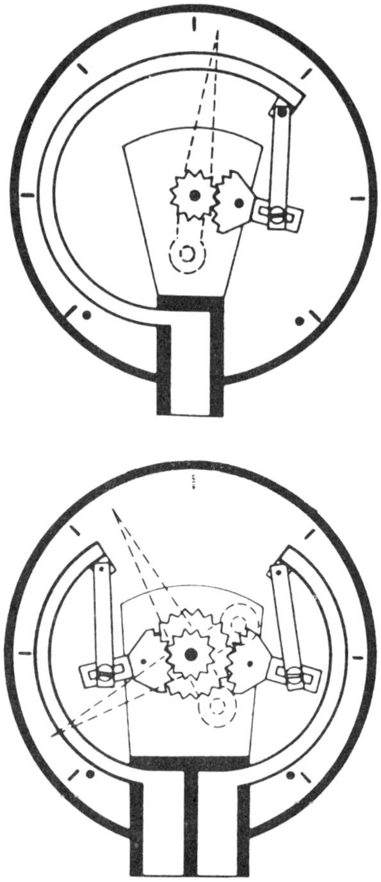

Figure 6.51 Air pressure gauges (WO)

the contacts touch, complete the circuit and cause the buzzer to sound. Each reservoir should have its own low-pressure warning device.

These devices are normally preset to operate at a pressure demanded by the design of the system.

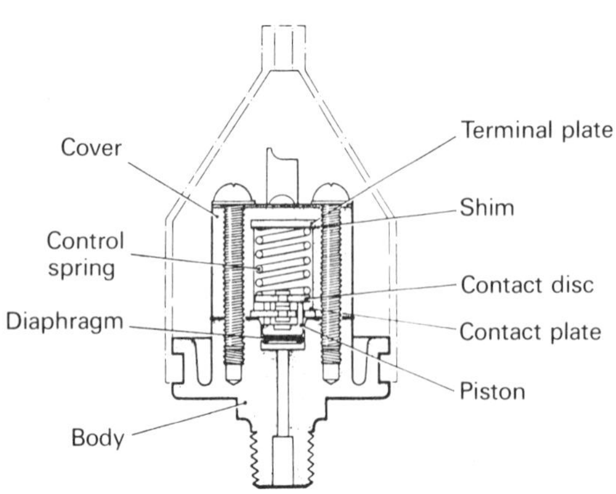

Figure 6.52 Low-pressure switch (BW)

Test point (CD)

EEC Directives require pressure test connections that comply with ISO 3583:1984 (see pages 5–6), to be fitted at specified points, for example both upstream and downstream of a load-sensing valve.

As shown in Figure 6.53, a threaded plastic dust cap (a) remains attached to this unit at (b) whilst the vehicle

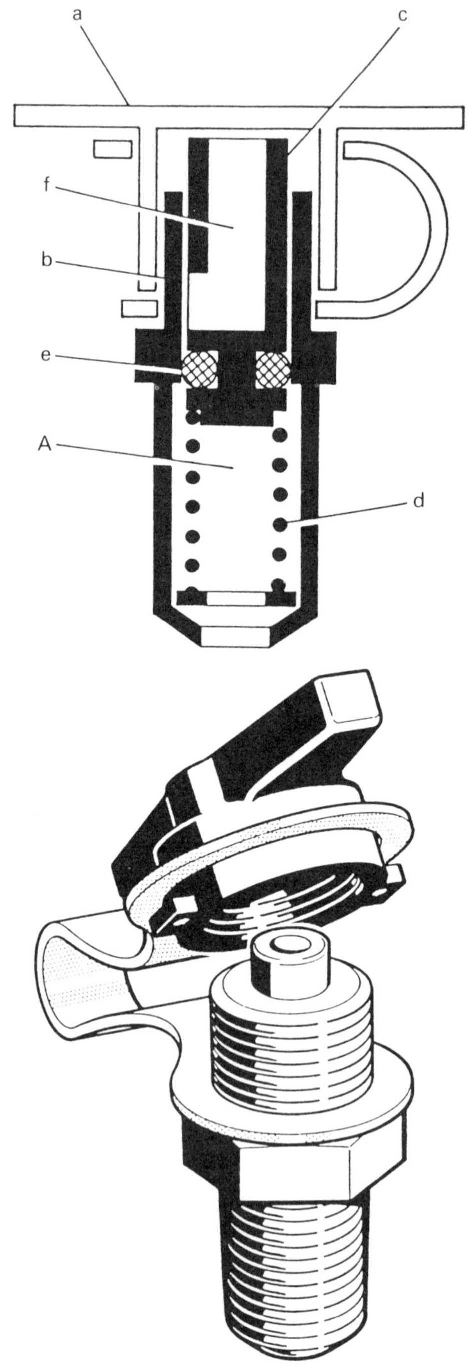

Figure 6.53 Test point (CD)

is in service, and only needs to be removed when a test gauge is attached via a suitable connector. As the threaded connector is affixed at point (b) it forces piston (c) down, against the resistance of spring (d), unseating seal (e) and providing an airway from chamber (A) to channel (f), and thus into the pressure gauge. When the connector is removed, spring (d) returns piston (c) to its original position, closing the airway.

Charging valves

As shown in Figures 6.54 to 6.56, compressed air enters the charging valve through port (g), under diaphragm (d), which is lifted against the restraint of spring (b) and piston (c). This opens port (e) and allows the air to continue en route to the reservoirs.

Charging valve with return flow (WO)

As shown in Figure 6.54, this valve, fitted between two storage reservoirs, allows the air pressure in the first reservoir to build up to its correct working pressure, before charging the second reservoir. If pressure drops in the first reservoir by more than 1.5 lbf/in^2 (10 kN/m^2, 0.1 bar) air flows back from the second reservoir, opening check valve (f) in the charging valve.

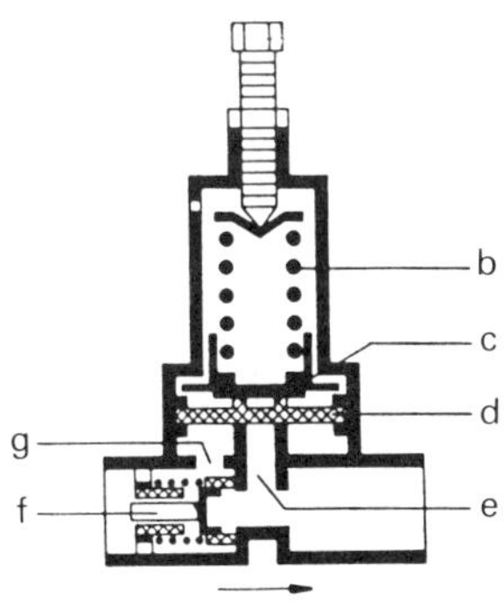

Figure 6.54 Charging valve with return flow (WO)

Charging valve without return flow (WO)

As shown in Figure 6.55, this valve is fitted to allow air pressure to build up in brake reservoirs, before allowing pressure to pass to auxiliary equipment, for example an air-operated clutch, doors or horns.

A return flow is not possible, because non-return valve (h) is kept closed by the higher pressure in the second reservoir.

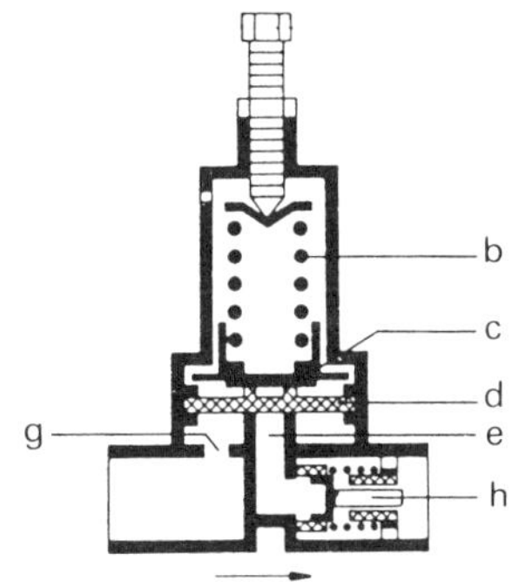

Figure 6.55 Charging valve without return flow (WO)

Charging valve with limited return flow (WO)

As shown in Figure 6.56, this valve is fitted to allow air pressure to build up in the brake reservoirs before allowing pressure to pass to the trailer, auxiliaries or parking system, etc. It can also be used as a pressure-protection device for a tractor. In the event of a failure of the emergency (supply) line to the trailer, when pressure drops in the tractor reservoir there is a partial return flow of compressed air until the closing pressure is reached.

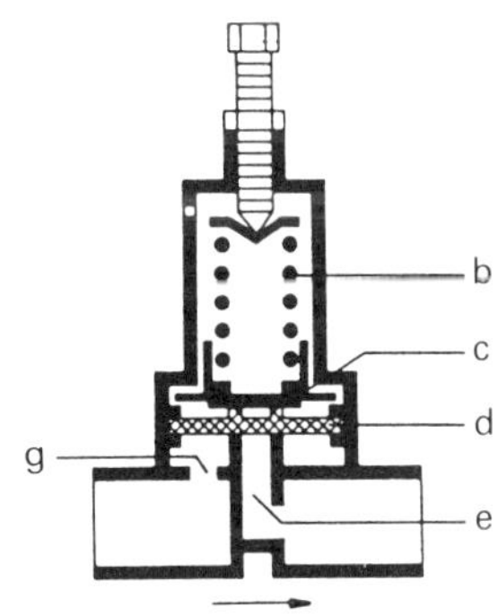

Figure 6.56 Charging valve with limited return flow (WO)

With the limited return flow version, air can flow back until the closing pressure of diaphragm (d) is reached, at which stage, spring (b) and piston (c) will reseat the diaphragm and prevent any further return flow.

Pressure-regulator valve and diverter valve (CD)

The pressure-regulator valve (see Figure 6.57) is fitted into the pipeline of a braking system to ensure that the system prior to this valve retains sufficient pressure,

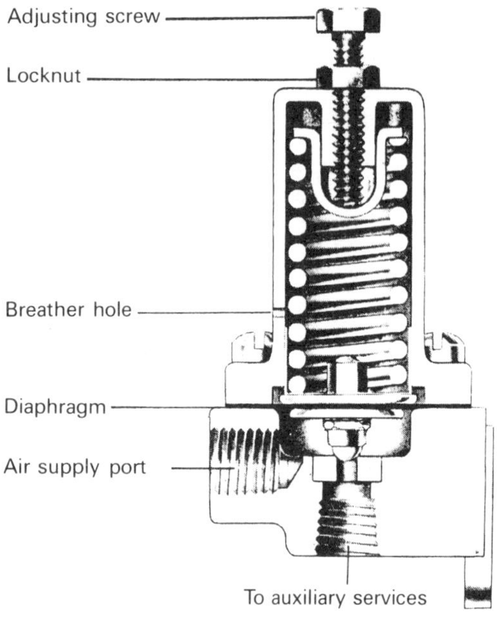

Figure 6.57 Pressure-regulator valve (CD)

even if this is to the detriment of the system beyond the valve. A typical function of this valve is to safeguard a braking system against the loss of air due to the excessive use of air by auxiliaries, such as an air-operated clutch, windscreen wipers, doors, etc. For this reason, it is often fitted between two reservoirs, to ensure that the first reservoir builds up a working pressure before the second reservoir is charged with compressed air.

When this type of valve is fitted directly to a reservoir, it has a different body and is known as a *diverter valve*. This type of valve is sometimes fitted with a non-return valve in the supply port, to feed air back to the main air system if the pressure in the auxiliary reservoir exceeds that in the brake reservoir, as could happen if the compressor should cease to operate.

The basic operation of either of the above valves relies upon a spring-loaded diaphragm, set to the recommended pressure for the particular vehicle, which lifts when that pressure is achieved, allowing the compressed air to progress further, as long as the high pressure is maintained.

Some pressure regulator valves have the supply port at the side and the outlet to auxiliaries at the base; whilst on others the supply is at the base and the outlet at the side of the valve. In either case, it is usual to find arrows on the casing of the valve to show the correct direction of the air flow.

Pressure-protection valve (BW)

The purpose of this valve is to prevent the flow of compressed air beyond the part of the system where the valve is fitted and thereby safeguard critical sections of the braking system from losses in pressure that may be suffered in other parts of that system. When fitted between two reservoirs it ensures that the first tank charges quickly, for braking use, whereas the next tank may be for a secondary circuit or for auxiliary equipment.

This valve is sometimes fitted in the emergency line of a tractor, so that losses on the trailer do not affect the tractor's brakes. In these circumstances, the trailer would also be protected by the relay emergency valve that is normally fitted to trailers, which automatically applies the trailer brakes in the event of a severe pressure loss in the emergency (supply) line.

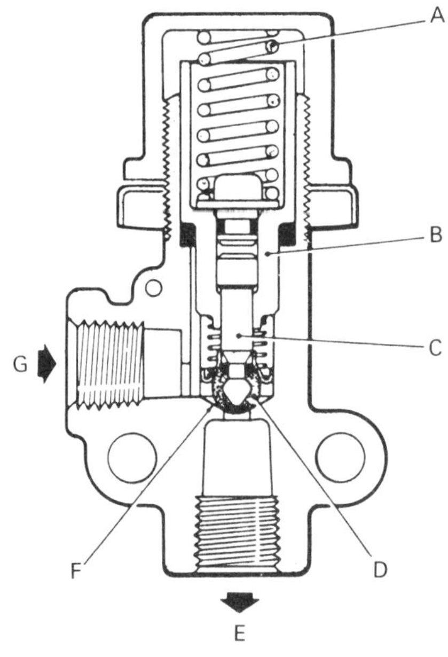

Figure 6.58 Pressure-protection valve (BW)

As shown in Figure 6.58, compressed air is supplied to inlet port (G) and the pressure builds up beneath piston (B) until it overcomes the force of the control spring (A). The piston then rises, lifting valve stem (C), thus opening the inlet valve (D) at (F), allowing the air to pass through delivery port (E). When the air pressure below the piston drops to such a degree that it fails to overcome the force of the control spring, that spring causes the piston to move down and prevent air passing through the valve.

One type of pressure-protection valve looks similar to the pressure-reducing valve (RV1) which is discussed below. The pressure-protection valve may be identified by its white cap, as opposed to the black cap of the pressure-reducing valve (RV1). In situ, the difference between the two valves is that the pressure-protection valve supply port is on the side of the unit and the delivery port at the base, whereas the pressure-reducing valve (RV1) is connected in the reverse fashion. Off the vehicle, a valve can be seen inside the bottom port of the pressure-reducing valve (RV1) but this is not the case with the pressure-protection valve. (**Note:** This does not apply to all types of these two valves.)

Pressure-reducing valve (RV1) (BW)

This unit is used in the system to reduce the pressure and maintain it at a constant pressure for ancillary circuits.

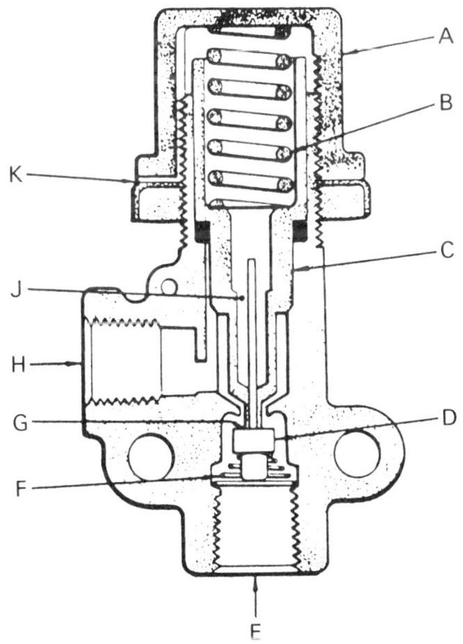

Figure 6.59 Pressure-reducing valve (RV1) (BW)

As shown in Figure 6.59, until air pressure builds up at port (E), the spring (B) holds down the piston (C) and the valve (D), closing the exhaust at (G), and opening the inlet.

Air can now flow from the supply port (E) past the open valve and out of the delivery port (H) until the pressure under piston (C) overcomes the force of spring (B). This allows the valve return spring (F) to close the inlet. When pressure drops in the delivery line the force of the control spring overcomes the reduced pressure under the piston, moving it down, opening the inlet valve and allowing the air pressure to again build up to the set level.

If the pressure in the delivery line exceeds the setting of the valve, the air pressure under the piston will overcome the force of spring (B) and the piston will rise, opening the exhaust at (G), allowing air to escape to atmosphere through the hollow piston at (J) and the exhaust vents (K) in the adjusting cap (A), until the preset pressure is reached.

Dual foot valve

Dual foot valves are now in general use, as by splitting the valve into two separate units, an additional safety factor is added to the service system. If one half fails the remaining half will function correctly, although there will then be a reduction in the overall braking efficiency of the service system. There are two basic types, *delta* (two valves side-by-side) and *concentric* (one above the other) and variations of each type.

There are several methods of connecting a dual valve into the braking systems of a vehicle. One typical layout is to connect one reservoir to one inlet port of the valve and another reservoir to the other inlet port. Each half of the valve has a delivery port to supply half of the braking circuit. By this method, if half of the system fails, the other half of the foot valve should continue to function correctly, acting upon the other half of the system's brakes.

Another method that can be used, is for the concentric type foot valve to have the upper inlet port connected to one reservoir and the corresponding delivery port on that valve connected to the *whole* of the service brake circuit. The lower inlet port is connected to a separate reservoir and the corresponding delivery port is connected to the *secondary* ports on double or triple-diaphragm brake chambers. A biased (stronger) spring is fitted to the lower piston and as a result, under normal circumstances, a driver obtains full service braking by only a light application on his treadle. Thus, only the top half of the foot valve is in regular use and it may well be that the driver does not even realise that the treadle will go down further than it does for conventional driving. If the service brake should fail, instinct would cause the driver to push hard down on the treadle in an effort to make the brakes operate and this effort would overcome the heavy-duty spring, so causing the lower piston to operate the *secondary* braking circuit.

Note: This type of system does not require a separate hand control for the secondary braking system.

Dual concentric foot valve (CD)

As shown in Figure 6.60, the main components of this type of dual concentric foot valve are an upper piston, a lower piston, a valve carrier piston positioned between them, and also two spring-loaded inlet/exhaust valves (one in the body and one in the valve carrier piston). There is a graduating spring at the top of the unit and four various springs interspaced between components.

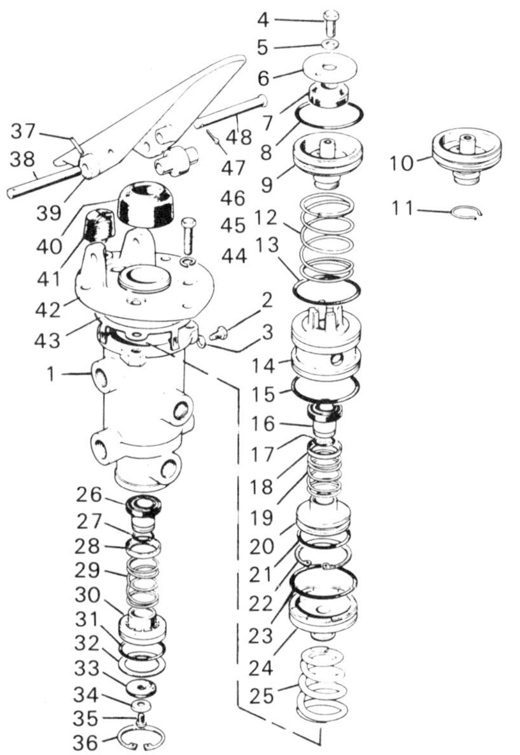

Key

1 = Body
2 = Stop bolt
3 = Washer
4 = Setscrew
5 = Washer
6 = Washer
7 = Rubber spring
8 = Sealing ring
9 = Piston
10 = Piston (later)
11 = Spring ring (later)
12 = Spring
13 = Sealing ring
14 = Valve carrier
15 = Sealing ring
16 = Inlet/Exhaust valve
17 = Sealing ring
18 = Valve retainer
19 = Spring
20 = Valve guide
21 = Sealing ring
22 = Circlip
23 = Sealing ring
24 = Piston
25 = Biased spring
26 = Inlet/Exhaust valve
27 = Sealing ring
28 = Valve retainer
29 = Valve spring
30 = Valve guide
31 = Sealing ring
32 = Washer
33 = Diaphragm
34 = Diaphragm washer
35 = Screw
36 = Circlip
37 = Tension pin
38 = Hinge pin
39 = Treadle plate
40 = Gaiter
41 = Treadle stop
42 = Mounting plate
43 = Spring clip
44 = Setscrew
45 = Washer
46 = Roller
47 = Split pin
48 = Fulcrum pin

Figure 6.60 Dual concentric foot valve (CD)

When the brakes are off the spring-loaded inlet/exhaust valves are seated with the inlet closed and the exhaust open, so that any air pressure in the brake chambers can exhaust to atmosphere.

When the treadle is depressed, the force is transmitted via the graduating spring to the pistons, and the first movement brings the hollow pistons into contact with the inlet/exhaust valves, closing the exhaust valves of both pistons, and sealing the brake chambers from atmosphere. Further downward movement of the pistons unseats the inlet/exhaust valves, and allows reservoir pressure to pass through the valve to the brake chambers.

Whatever air pressure is present within the brake chambers, is also present in the cavities beneath the upper and lower pistons. When this pressure below the pistons matches the mechanical force exerted downwards by the graduating spring, all three pistons lift and the inlet valves close, cutting off further compressed air supply to the brake chambers. The valve is then in the balanced position, with equal pressure above and below the pistons and with equal pressures in both front and rear brake chambers. If the treadle is further depressed by the driver, the pistons will all move down and the inlet valve will again be unseated, allowing extra pressure to pass to the brake chambers, until a balance is again reached, when the inlet valves will close again. When the treadle is released, the inlet valves close as the pistons rise and then the exhaust valve opens, restoring atmospheric pressure to the brake chambers.

Dual foot valve (delta) (BW)

This type of foot valve can be considered as being similar to two single foot valves which are connected together, one behind the other, and operated by a link, so that they operate simultaneously as well as equally whenever the treadle is used by the driver. As with other dual valves, the failure of one half of the valve or one half of the service system will not prevent the other half from working.

As shown in Figure 6.61, when the brake treadle is applied, link (A) acts down on to plunger (B), causing spring retainer (C) to compress graduating spring (D). This spring moves down piston (E) which contacts

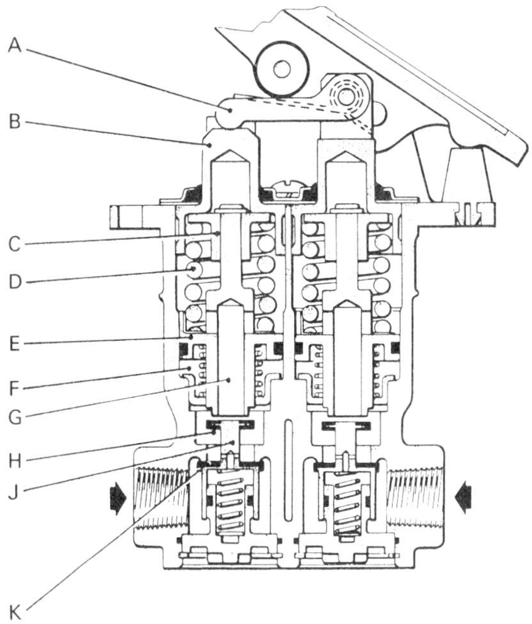

Figure 6.61 Dual foot valve (delta) (BW)

inlet/exhaust valve (J) and closes the exhaust outlet (G). Further downward movement of piston (E) opens inlet valve (K) and allows air from the service reservoir to flow to the brake chambers via supply port (H).

Note: The delivery ports of this valve are at 90° to the section drawing at point (H) and therefore are not shown.

Whilst air is being delivered to the brake chambers, it is also present in chamber (F), below piston (E). When the air below the piston overcomes the downward force of the graduating spring above, the piston will rise, closing the inlet valve and also leaving the exhaust closed. At this stage, with the force above and the pressure below the piston equal, this is known as the *lapped* (or *balanced*) *condition*. If the driver increases his force on the treadle, the piston will move down again, opening the inlet valve so that additional pressure can be passed to the brake chambers. The pressure can then build up until a balanced position is again achieved. When the pedal is finally released completely, air passes through exhaust outlet (G) and escapes to atmosphere via a shared exhaust port (not shown on this figure) in the centre of this unit at 90° to this figure, above the delivery ports.

E6 foot valve (BW)

As shown in Figure 6.62, this is a dual foot valve containing two elements, one positioned above the other, where each element operates an independent brake circuit and is unaffected by a failure of the other element or circuit. When the footbrake treadle is applied, this mechanical force acts upon spring seat (Q) and rubber graduating spring (R), causing piston (N) to move down, and for exhaust valve seat (M) to close the exhaust route to (F). As the downward movement of piston (N) continues, inlet valve (K) is moved off its seat (L), allowing compressed air from No. 1 supply port to flow through the foot valve and leave by No. 1 delivery port, en route to the brake chambers in that circuit.

The downward movement of upper piston (N) also moves down spring (G) and lower piston (E), which causes exhaust seat (D) to close off the hollow exhaust route (F). As air passes through No. 1 delivery port, it also passes through transfer holes (J) and acts on the top of piston (E), causing it to descend and to move inlet valve (B) off its seat (C). Compressed air from No. 2 supply port is then able to flow through the foot valve and leave by No. 2 delivery port, en route to the brake chambers in that circuit.

Lapped or balanced condition

It is essential that the design of a control valve relates the pressure of the air supplied to brake chambers, to the force applied by the driver to the footbrake treadle. As shown in Figure 6.62, after the application of the brake treadle, air continues to flow out of No. 1 delivery port and the pressure continues to rise, until the air pressure beneath piston (N), together with the force from return spring (O) and the return spring beneath inlet valve (K), all acting upwards, exceed the force acting down on piston (N). It will then lift slightly, and allow inlet valve (L) to be closed by its return spring. The exhaust valve (M) remains closed. Thus, both the inlet valve and the exhaust valve of No. 1 circuit are closed, but leave the brakes applied. This is known as the *lapped* or *balanced condition*.

If extra force is then applied to the foot treadle, piston (N) is moved down and, as previously, causes the inlet valve to open, allowing the air pressure in the system to increase until the forces above and below the piston are equal.

If the force on the treadle is partially released, piston (N) will rise, leaving the inlet valve closed, but temporarily opening the exhaust route, until the forces above and below the piston are equal.

A similar situation occurs in the other element of this foot valve. As air continues to flow out of No. 2 delivery port, tell-tale pressure acts beneath piston (E), and this together with the force of the return spring within valve

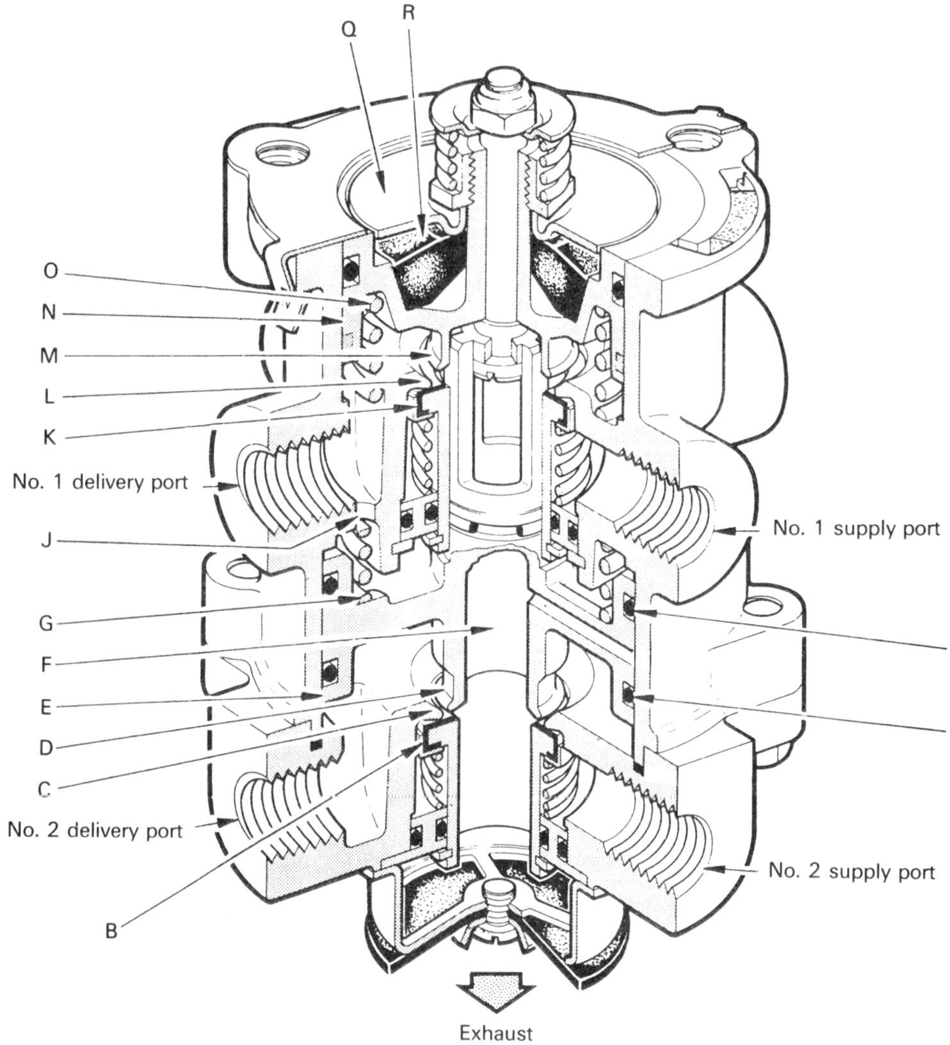

Figure 6.62 E6 foot valve (BW)

(B), overcomes the downward force of spring (G) and the air pressure above (E). Inlet and exhaust valves are closed, but the brakes are left applied.

In the event of air pressure being lost in No. 2 circuit, No. 1 circuit will continue to operate on the brakes to which it is connected, and seals (V) will prevent cross-leakage.

Should No. 1 circuit lose air pressure, No. 2 circuit will continue to operate on the brakes to which it is connected, as a result of piston (N) contacting piston (E) which closes exhaust seat (D) and opens inlet valve (B), allowing air to flow between supply port No. 2 and delivery port No. 2.

Dual foot valve with electric switch (WO)

To avoid a repetitive description of the internal operation of the two identical elements of this foot valve, the following paragraphs refer to both of these items.

In addition to having two independent valve units to control the air brakes, this particular foot valve incorporates an electrical switch to operate the exhaust brake.

As shown in Figure 6.63, as treadle (1) is applied, plunger (2) is moved downwards on to cup (4), so compressing spring (5). Simultaneously, the same action is taking place in the neighbouring element.

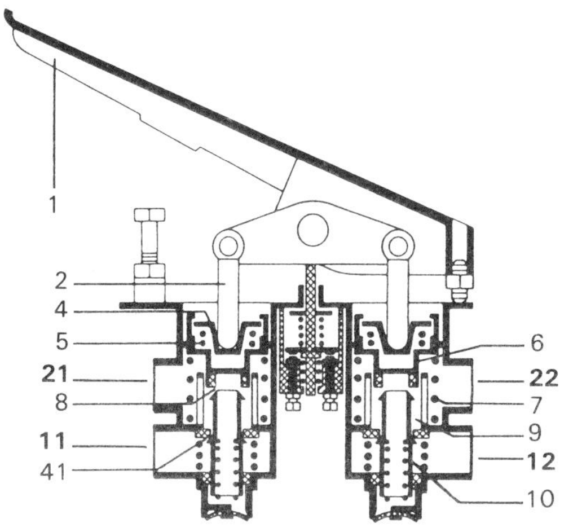

Figure 6.63 Dual foot valve with electrical switch (WO)

Thus, piston (6) is moved down against spring (7), closing exhaust valve (8). Inlet valve (41) is opened by the downward movement of hollow stem (10). Compressed air from the reservoir enters chamber (9) via port (11), en route to the brake chambers, via port (21).

Air pressure, from the reservoir, builds up in chamber (9). When that pressure exceeds the downward force applied, piston (6) will move up, against the force of spring (5), resulting in the inlet valve closing. Air cannot escape from the brake chambers, and thus the foot valve is in the lapped (or balanced) condition. Naturally, if the driver either increases or decreases the force on the foot valve treadle, the valve will respond as necessary, i.e. by increasing or decreasing the air pressure supplied to the brake chambers.

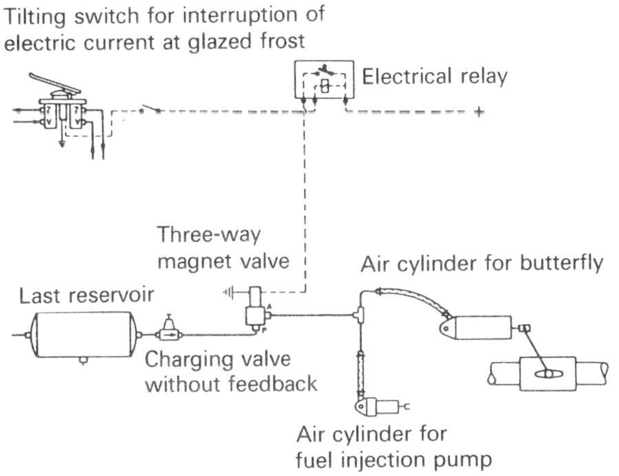

Figure 6.64 Layout for exhaust brake system

This particular foot valve assembly is designed so that when the treadle is pressed, an integral electrical switch is closed, causing a three-way magnet valve to operate the exhaust brake, *before* air begins to flow to the brake chambers. This can be seen in Figure 6.64.

Note: There is a variation of this foot valve, where instead of an electrical switch, another air valve is operated. This too activates the exhaust brake.

Three-way magnet valve (WO)

This valve can, by an electrical signal, cause an air line to be charged or exhausted.

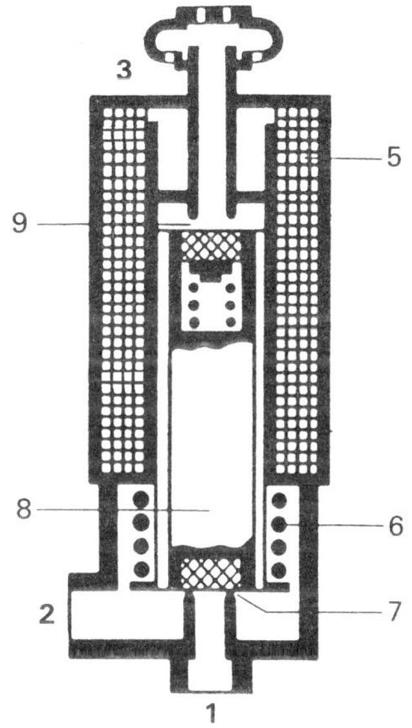

Figure 6.65 Three-way magnet valve (WO)

As shown in Figure 6.65, port (1) is connected to a reservoir. Return spring (6) holds inlet valve (7) at the lower end of solenoid core (8), against its seat. When solenoid coil (5) is de-energised, spring (6) returns solenoid core (8) and inlet valve (7) to the closed position. This action automatically opens exhaust valve (9) and air escapes to atmosphere via exhaust vent (3).

Note: A capacitor is connected in parallel to the solenoid coil to eliminate arcing of the contact breaker switch.

Air cylinder (WO)

The role of these units is to shut-off the fuel injection pump and to close the butterfly valve in the exhaust brake.

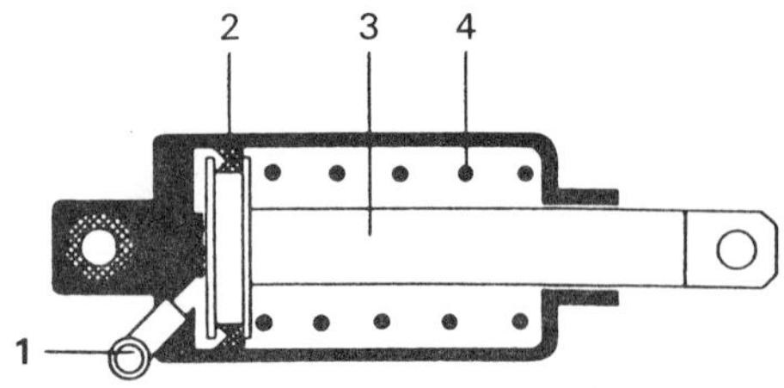

Figure 6.66 Air cylinder (WO)

As shown in Figure 6.66, compressed air from the three-way valve enters at port (1) and acts against piston (2), push-rod (3) and return spring (4).

The cylinder shown is connected to a lever on the fuel injection pump, so that the lever can be moved from the idle to the stop position, when the exhaust brake is used. There is another connection to the throttle linkage to prevent the accelerator pedal being depressed, whilst this brake is in use.

Another cylinder, a variation of Figure 6.66, is connected to the butterfly in the exhaust system. When operated the butterfly closes and the resultant back pressure gives engine braking.

Three-way valve (WO)

The role of this mechanically-activated valve, is to charge and discharge the air cylinders that control the exhaust brake.

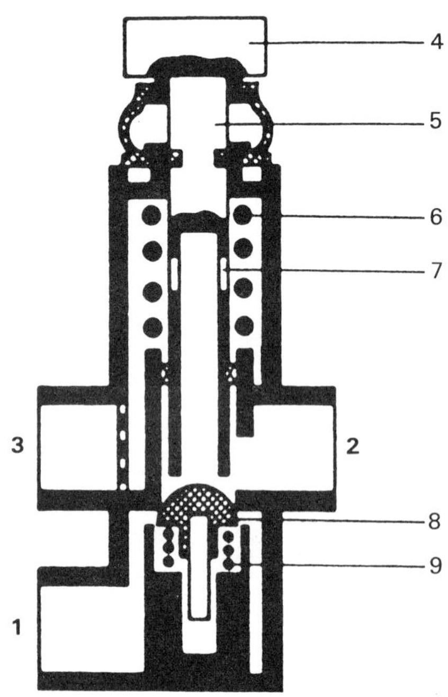

Figure 6.67 Three-way valve (WO)

As shown in Figure 6.67, air from the reservoir enters via port (1) and acts beneath inlet valve (8). Thus, as plunger (4) is pushed down, stem (5) unseats inlet valve (8), allowing compressed air to pass to the air cylinders via port (2). When plunger (4) is released, spring (6) returns stem (5) to TDC. Spring (9) closes inlet valve (8).

The air cylinders exhaust to atmosphere via drillings (7) and exhaust port (3).

Exhaust brake (WO)

This is one of several alternative additional retarder systems. It is not intended to bring the vehicle to rest, but to supplement the orthodox braking systems, during normal journeys. Even though the use of low gears on down gradients is an accepted and necessary safe method of driving heavy vehicles, it still requires prolonged use of the braking system. This can cause severe overheating of brake linings and drums, resulting in brake fade. An exhaust brake shuts off the exhaust system, causing back pressure and giving engine braking. It does leave the drums and linings cool and it does increase the mileage before relining is necessary. In some countries, for example, Germany, the fitting of this device to certain vehicles is a legal requirement.

A layout suitable for buses that operate in heavy traffic is shown in Figure 6.64.

An application of the orthodox foot valve causes an electrical relay and a three-way magnet valve to operate. Therefore, whenever a service brake application is made the exhaust brake is switched on.

In the case of an HGV and trailer, the exhaust brake works as described above, although some systems in addition have the trailer brake activated whenever the exhaust brake is used.

Although an electrical device has been described for operating the exhaust brake, a mechanically-operated valve, as shown in Figure 6.67, is in common use. It is often operated by a floor-mounted button.

A butterfly valve is shown in Figure 6.64 but there are alternatives to this and these include the sliding-gate type, which is claimed to be self-cleaning.

Stop-light switch

Stop lights must be fitted to vehicles in such a way that they operate when either the service or secondary brakes are applied. The modern switch is so sensitive to pressure that a mere 5 lbf/in^2 (34 kN/m^2, 0.3 bar) will cause the brake lights to function. Therefore, even a slight application of either the foot or hand valve will operate the stop lights.

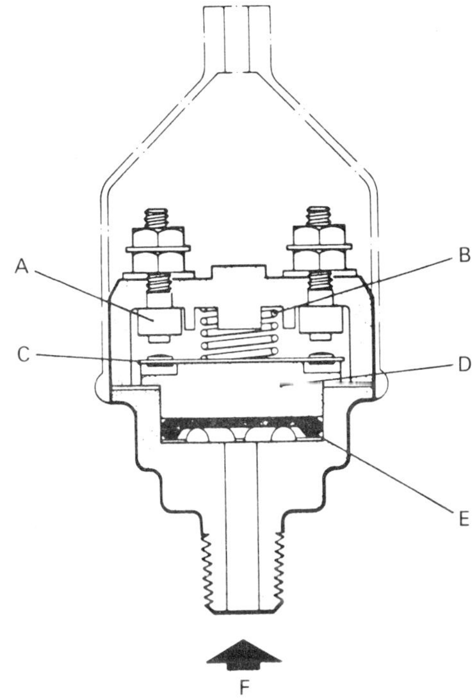

Figure 6.68 Stop-light switch (WO)

As shown in Figure 6.68, the switch body contains a diaphragm/seal (E) which moves piston (D) upwards in response to compressed air entering port (F). The top of the piston will compress return spring (B), so allowing contact (C) to bridge terminals (A) completing the electrical circuit and lighting the stop-lamp bulbs. When the brakes are released, the pressure escapes from the switch and the return spring moves the contact strip away from the terminals, breaking the electrical circuit.

On the secondary circuit of a spring-brake system, where air pressure is reduced to apply the brakes, a low-pressure switch is used. This type of switch uses the reverse principle to the normal stop-light switch, i.e. when the brakes are off air pressure holds electrical contacts apart, but as the air pressure is reduced, the return spring brings the contacts together, completing the circuit and lighting the stop-light bulb.

Brake actuators

There are many types of brake chamber produced including: single-diaphragm; double-diaphragm; double-diaphragm combining spring brake and service brake; piston/diaphragm, with piston spring-brake and diaphragm service-brake; double piston, for spring and service-brake; remote piston; as well as remote diaphragm spring-brake chambers.

It is of the utmost importance that only the recommended type and model of replacement brake chamber is used.

Single-diaphragm brake chambers (BW)

As shown in Figure 6.69, the spring chamber consists of a pressure plate and a non-pressure plate containing a spring-loaded push-rod assembly and a diaphragm. The plates and diaphragm are clamped together, by a clamp ring, with bolts and nuts. The push-rod return spring is located in the non-pressure plate by a spring. This type of unit can also be fitted to operate a clutch or master cylinder.

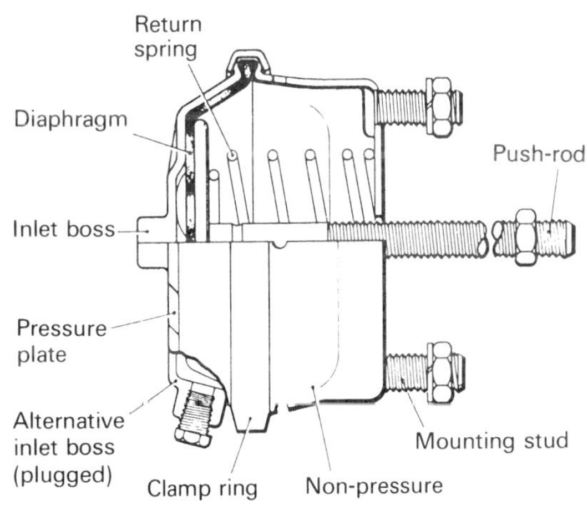

Figure 6.69 Single-diaphragm brake chamber (BW)

The single-diaphragm-type brake chamber, converts the energy of compressed air into the mechanical force necessary to operate the brakes. As air pressure enters into the chamber behind the diaphragm, the push-rod is moved to operate the camshaft and apply the brakes.

Release of the air pressure allows the mechanism to return to the 'off' position through the action of the return springs.

Double-diaphragm brake chamber (BW)

As shown in Figure 6.70, the double-diaphragm brake chamber is similar in construction to the single-diaphragm type (Figure 6.69). The construction is extremely simple and consists of a chamber fitted with two diaphragms - a service diaphragm, fed from a side

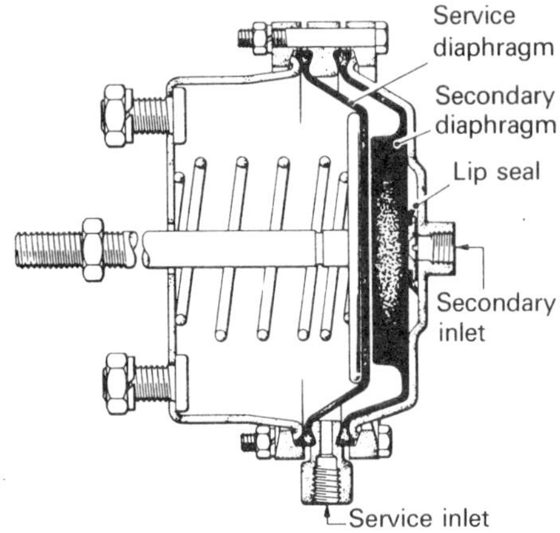

Figure 6.70 Double-diaphragm brake chamber (BW)

entry, and a secondary/emergency diaphragm fed from a central rear entry in the pressure plate.

A double-diaphragm brake chamber is designed to provide an air-braked vehicle with a compact and reliable brake operating device which combines the function of a service as well as a secondary/emergency/parking brake (the latter being in conjunction with an air-assisted handbrake). If one diaphragm fails, the other will still operate the brakes.

There is a small lip seal fitted to the rear of the secondary diaphragm covering the inlet port, so that should the rear diaphragm fail, service air will not escape from behind the service diaphragm into the secondary system.

Note: Over a period of time, air pressure is likely to leak, and parking can only be sensibly and legally achieved with this type of chamber when it is used in conjunction with an air-assisted handbrake lever. The mechanical linkage will hold the foundation brake applied by mechanical means and the air in the brake chamber is then vented to atmosphere.

Spring brakes

The name spring brakes is usually applied to the type of brake chamber that contains a piston or diaphragm, and a very powerful spring that is held in compression during normal driving, by compressed air. When the compressed air is released, the spring applies the brakes mechanically. This device can be used to apply the secondary or the parking system, by means of a graduable hand valve. This hand valve has three positions: 'off', secondary and park. Between secondary and 'off' positions, the driver can use the graduable control so that the brake can be progressively applied.

The spring brake also has an important safety aspect, for if there is a loss of pressure in the secondary/park circuit the springs will overcome the lower, or lost, pressure and apply the brakes automatically, not releasing them until the pressure is restored.

There are many different types of spring brakes, but they can be divided into groups as follows:

1 Those with tandem chambers, where one is a conventional single-diaphragm brake chamber that is operated by the foot valve, and the other chamber in the pair is a spring chamber (piston or diaphragm) which is connected to the hand valve of the secondary/park system. Such chambers have a mechanical wind-off device to release the brakes without the aid of air pressure.

2 Those that provide *only* secondary/park brakes. These units, known as remote spring brakes, consist only of the spring chamber and lack the addition of the conventional chamber as incorporated in the first type above. This second type usually operates by a pull action, whereas the first type usually operates by a push action (although there are exceptions to this rule).

This second type can be sub-divided into:

a) those with a full wind-off device;

b) those without a full wind-off device.

One such chamber is used to operate mechanical linkage to both brakes on the rear axle of a vehicle fitted with air/hydraulic service brakes.

Examples can be seen in Figures 6.71 and 6.72.

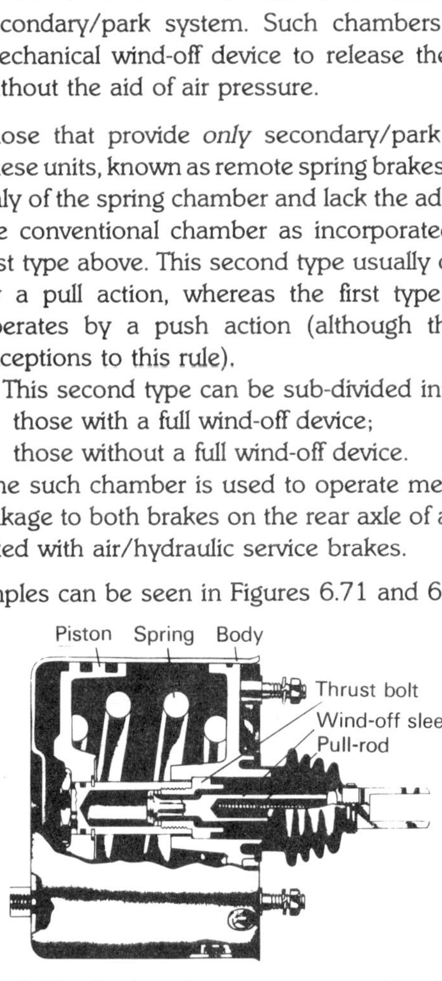

Figure 6.71 Section of remote actuator with wind-off device (CD)

With such a variety of units it is not practical to deal with every one. An example of the dismantling procedure of two of these units (a Bendix model and a Clayton Dewandre model) is given in Chapter 9. It is emphasised that these procedures are included to

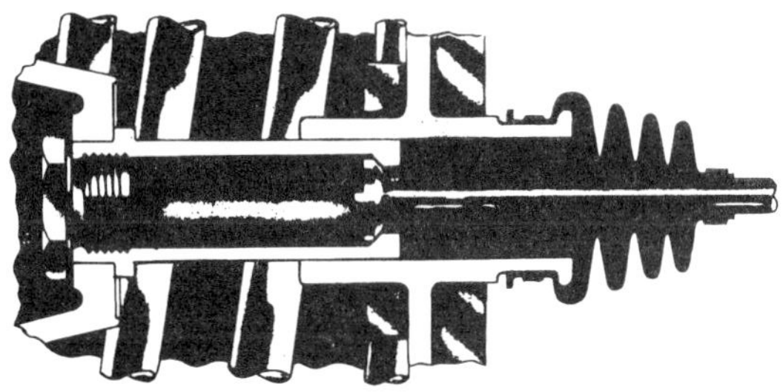

Figure 6.72 Partial section of remote actuator without wind-off device (CD)

stress the latent danger of the spring and although they are typical methods, the reader should check with the workshop manual of the particular vehicle to ensure that the correct method is used for the job in hand. However, the safety factor is common to all spring-brake chambers - do **not** attempt to dismantle any type of spring-brake chamber unless you have the specialised equipment and you are aware of the *particular* dismantling procedure.

To dismantle any of these units without adequate equipment or using the wrong method could result in serious or fatal injuries to the fitter or another person in the workshop. A compressed spring has a force of up to 3000 lbf (13 kN).

Spring brakes pose several problems. These include:

1 That in order to apply the brake, air has to be withdrawn from the brake chamber.
2 In the case of a tractor or towing vehicle, the conventional brake chambers on the trailer still require upright air when, simultaneously, the spring chambers on the towing vehicle need inverse air pressure.

These problems have been tackled in various ways. At one time Bendix circuits incorporated an inverse spring-brake valve, connected to the L3 hand valve. However, this arrangement has been superseded by the L4 hand valve, which can control both inverse and upright air and is merely linked to a relay valve. Clayton Dewandre Ltd produce a similar valve.

A straight-gate hand valve with a spring-loaded collar is now in common use with type-approved vehicles drawing a trailer. This control valve is mostly used in conjunction with a trailer control valve, which is capable of operating orthodox (upright air) brakes on the trailer.

Spring-brake chamber (BW)

As shown in the sectioned view of a spring-brake chamber in Figure 6.73, the two separate chambers are shown; service chamber (D) and spring-brake chamber (H). They operate independently from each other, but are fastened together for convenience of mounting and linkage.

The contents of the service chamber are the same as a standard, single-diaphragm brake chamber, i.e. push-

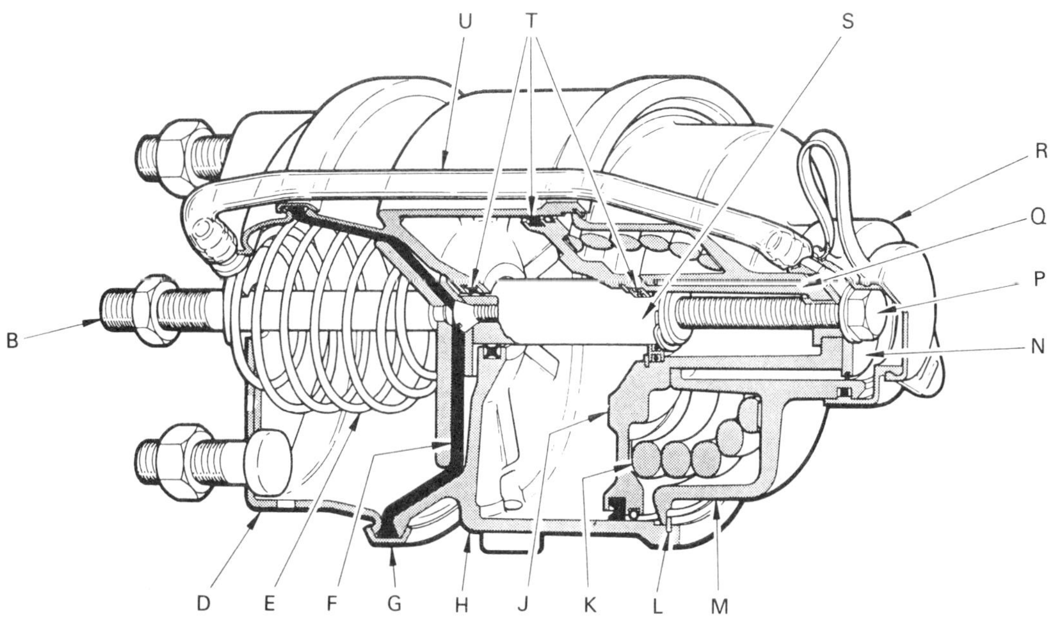

Figure 6.73 Spring-brake chamber (BW)

rod (B), return spring (E), diaphragm (F) and clamp ring (G).

The contents of the spring-brake chamber are basically the same as other types of spring-brake chambers, i.e. a piston (J) and powerful spring (K), together with large circlip (L) which holds the chamber head (M) in place. There is an anti-explosion washer (N) as a safety back-up, in the event of problems with the large circlip (L).

For normal running of the vehicle, the wind-off bolt (P) is left screwed into piston rod (S) and abutting against piston boss (Q). Air pressure seals for the piston and piston rod are shown at (T). Dust cap (R) is linked by a nylon breather pipe (U) to service brake chamber (D), to minimise the abrasive particles drawn into the spring-brake chamber.

The action of a double-chamber type with wind-off arrangements is shown in Figures 6.74 to 6.77.

Figure 6.73 shows the unit containing sufficient air pressure in the spring chamber to keep the spring compressed and thus the brakes in the 'off' position.

Figure 6.74 shows the service brake in use and the brakes applied. The spring remains compressed and inoperative, whilst the conventional brake chamber diaphragm is subject to air pressure, supplied via the foot valve, to operate the brakes in the orthodox manner.

Figure 6.75 shows that the air has been released from the spring chamber, as a result of the hand control valve being used. The first movements of the spring provide secondary braking and the rate of application of the brake is progressive and controllable. When all the air has been exhausted, the brake is held applied by the mechanical application of the spring. A leak in the secondary circuit, if not replenished by the compressor

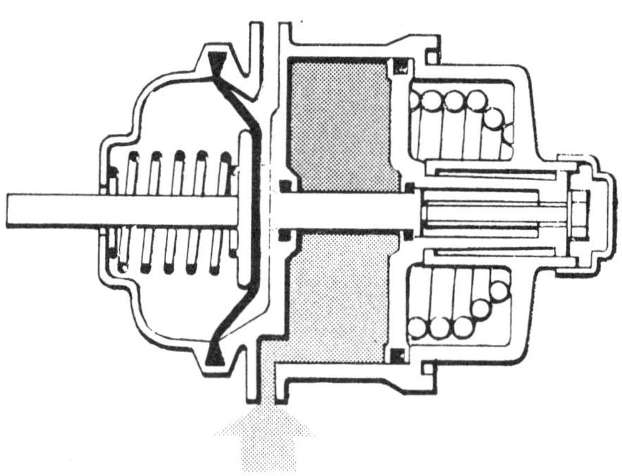

Figure 6.74 Brake off – normal driving

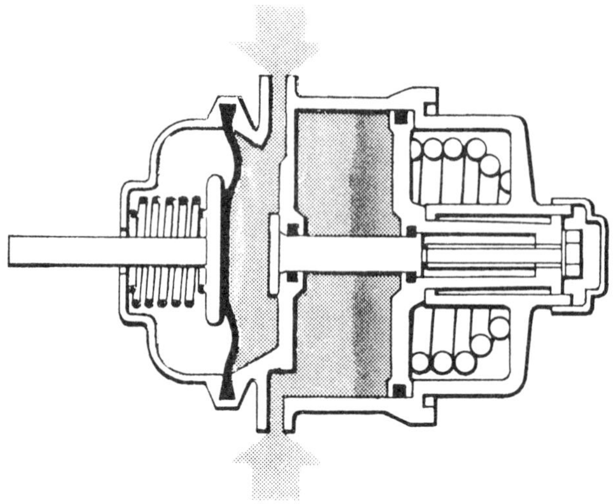

Figure 6.75 Service application – spring brake held released

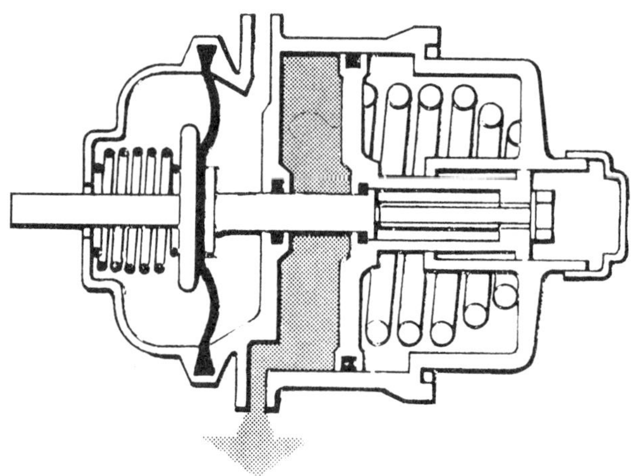

Figure 6.76 Secondary and parking – spring brake applied

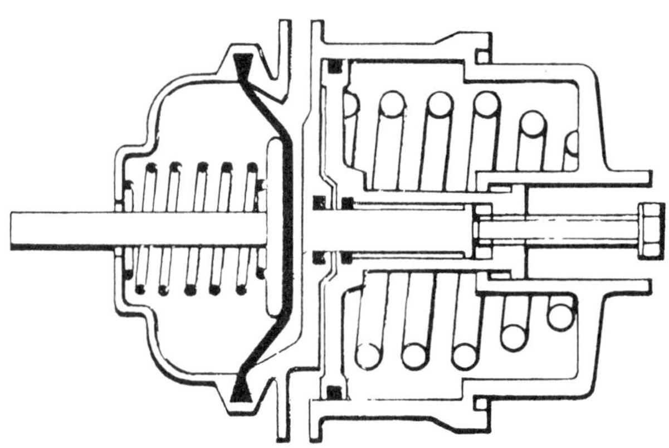

Figure 6.77 Manual release

will apply this brake. A leak in the service system of certain layouts can also activate the spring, by loss of pressure.

Figure 6.76 shows the action of the release bolt, to free the brake so that the vehicle can be moved, even without the aid of compressed air within the spring chamber. However, caution should be exercised in this action and the road wheels should be firmly chocked before the brakes are released.

Brake chambers for use with wedge-operated foundation brakes (BW)

Whereas the orthodox S-cam brake shoe expander usually has a shaft extending towards the centre of the vehicle, to which the push-rod from the brake chamber can be connected, a different layout is necessary with wedge-operated brakes.

As shown in Figure 6.78, the brake chamber is either bolted to the back plate of the foundation brakes or is tube mounted, being screwed directly into the body of the brake expander.

There are several variations of chamber in common use, ranging from single-diaphragm chambers, with or without a handbrake operating arrangement, to a service/secondary type of spring brake chamber.

Equally, the installation can be as varied, to allow for service, service/secondary/park arrangements and for leading/trailing or twin-leading shoes, as decided by the designer of the particular braking system.

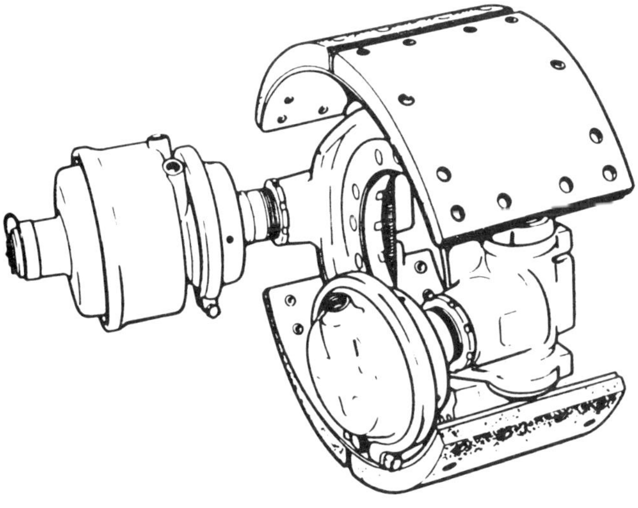

Figure 6.78 Brake chamber for wedge-operated brakes (BW)

Double-piston spring brake (KR)

The chamber closest to the connecting rod contains the secondary/park piston and the extremely powerful spring associated with spring brakes. The other chamber also contains a piston, but a lighter return spring.

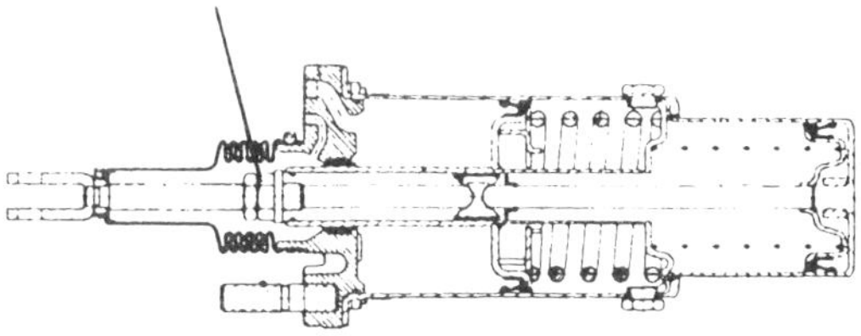

Figure 6.79 Double-piston spring brake (KR)

As shown in Figure 6.79, there is a release device (collar nut), which enables the brake to be wound off mechanically in the event or air loss, but dismantling of this unit requires a press.

Diaphragm spring brake (RR)

Although the well-known spring brake is a piston type, there is a large number of diaphragm spring brake units in current use. These are double-diaphragm chambers, where the front section is an orthodox service brake, whilst the rear is a secondary/park spring brake, which uses a diaphragm instead of a piston (see Figure 6.80).

Advantages of the diaphragm types are that they do not suffer from bore wear or O-ring failure, and are not likely to freeze up in severe conditions.

Reconditioning methods employed with these units are totally different from those used with the piston variety. A new complete spring chamber 'piggy-back' can be fitted whilst the brake chamber is still attached to the vehicle, without the use of a press or special tools (see the maintenance section for details).

Double-diaphragm brake chamber plus spring chamber (RR)

This is a variation of Figure 6.80, being a double-diaphragm brake chamber with the detachable spring chamber at the rear (see Figure 6.81). As with the previous model a new complete spring chamber 'piggy-back' can be fitted whilst the brake chamber is still attached to the vehicle, without the use of a press or special tools (see the maintenance section for details).

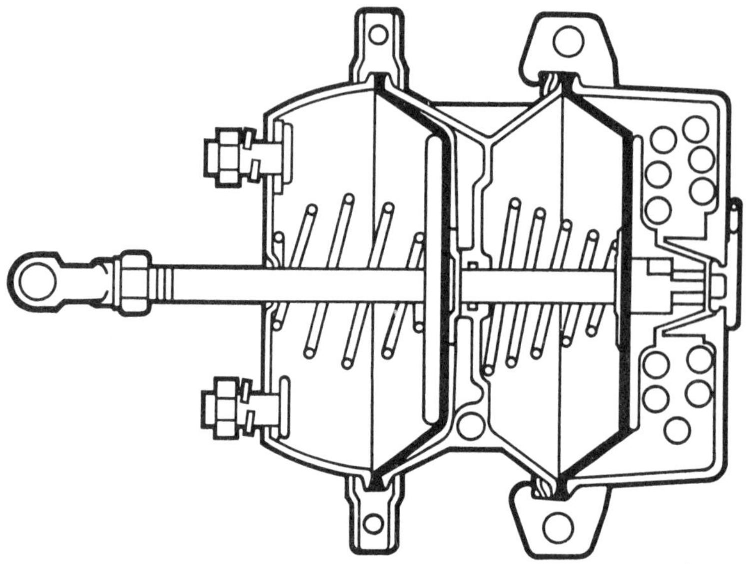

Figure 6.80 Diaphragm spring brake (RR)

Spring brakes are becoming more common on trailers in the UK, and with this arrangement both the tractor *and* the trailer can be held parked by the springs.

Safety is a major factor in the design of brakes and braking systems. The run-away vehicle, with the potential to cause a tragedy, has prompted the conscientious haulier to recheck his maintenance schedules and the manufacturer to strive constantly to find foolproof (and if possible neglect-proof) systems. The lock actuator, which was a step in the right direction, has been superseded by the spring brake, which with modern piping and valve arrangements can provide a positive fail/safe layout.

Another aspect of spring brakes is the fact that if air pressure and the spring pressure are *both* applied together, the spring-brake push-rod, the slack adjuster and the foundation brake will be either strained or damaged. To avoid this compounding of forces, most manufacturers employ some form of anti-compound valve, although the design differs. Bendix Ltd use differential-protection valves, whereas other companies use double-check valves (sometimes called change-over valves) or overload-protection valves.

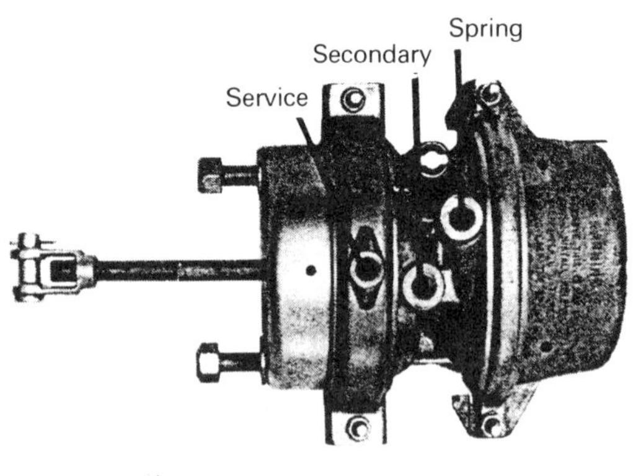

Figure 6.81 Double-diaphragm brake chamber plus spring chamber (RR)

Air-assisted handbrakes

It is a legal requirement that a parking brake must be held applied by mechanical means, for example by a rod/cable, spring brake or lock actuator. However, this does not prevent the use of air assistance to aid the driver in applying a mechanically-operated brake. Once the brake has been applied, with assistance, and then the air is released, that vehicle is held parked in the same manner as a vehicle that lacks air assistance.

There are many types of handbrake valves and many different mechanical systems for activating such valves. Figures 6.82 and 6.83 show one such valve and one mechanical layout.

As shown in Figure 6.82, the valve body contains a spring-loaded inlet/exhaust valve, above which is a spring-loaded hollow plunger and above this component is an operating plunger with an adjusting screw, to give the correct clearance from the handbrake mechanism.

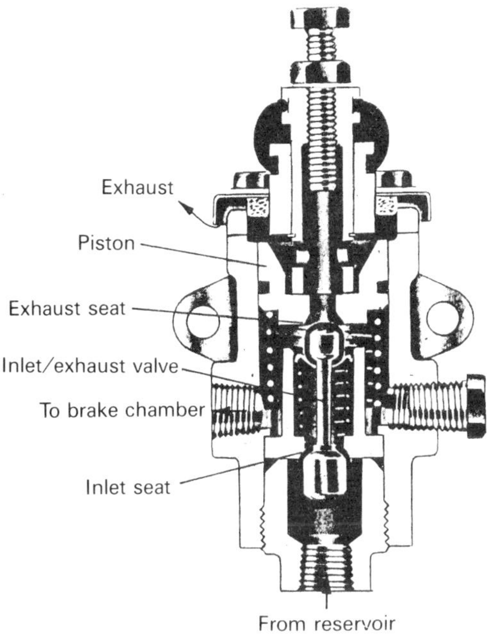

Figure 6.82 *Air-assisted handbrake valve (BW)*

As shown in Figure 6.83, when the handbrake lever is applied a mechanical force is applied, by the lower section of the lever, onto the top of the adjusting screw of the valve. This causes the operating plunger to move downwards, causing the hollow piston beneath, to move down and make contact with the inlet/exhaust valve. Thus the exhaust seat on the hollow plunger is sealed off, and further downwards movement by the plunger unseats the inlet/exhaust valve from the inlet seat, allowing compressed air from the reservoir to enter the valve. This air then leaves the delivery port and passes to the brake chambers where it applies the brakes.

When the driver releases his hold upon the handbrake lever, to park the vehicle, a spring-loaded pawl on the power lever engages with the ratchet on the handbrake lever, to hold the brakes applied. The spring reaction of the handbrake air valve, plus the reaction of the linkage caused by the action of the brake chamber, causes the handbrake lever to move slightly about its pivot pin, towards the 'off' position, closing the inlet port and opening the exhaust port, so that the pressure within the brake chamber is exhausted to atmosphere leaving the handbrake held 'on' by mechanical linkage.

To release the handbrake, the driver pulls the handbrake lever, which reacts against the air-assistance valve, so that compressed air from the reservoir is supplied to the brake chambers. When the ratchet lever is squeezed against the handbrake lever, the pawl lifts, the lever can then be moved to the 'off' position, sealing the inlet valve, but allowing the air pressure in

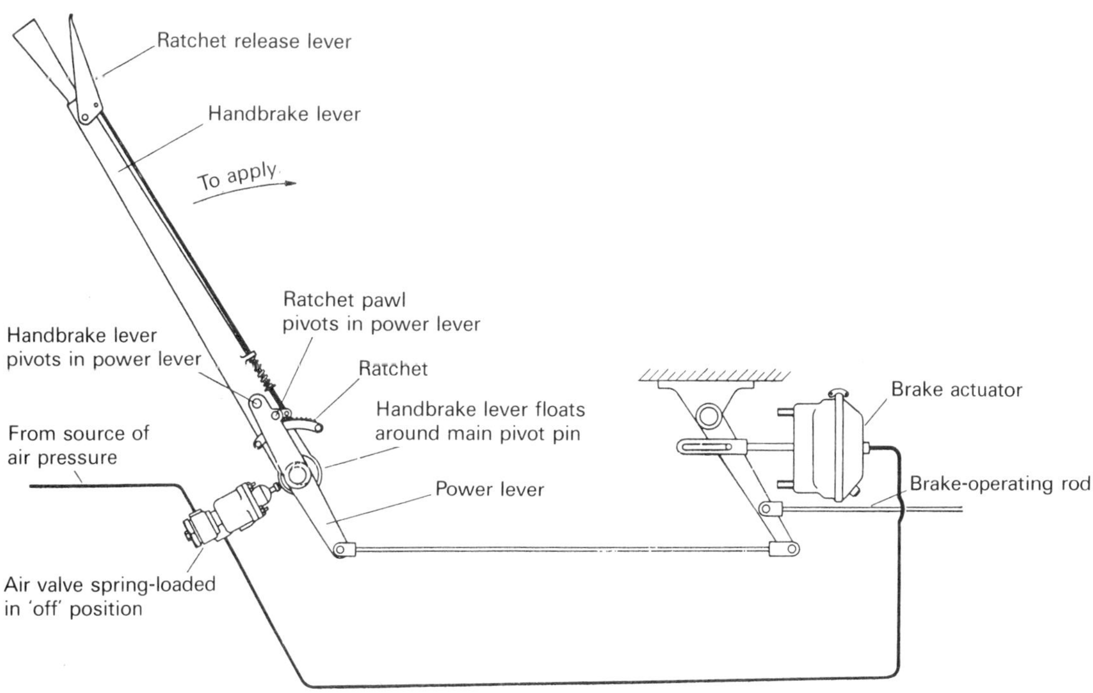

Figure 6.83 *Air-assisted handbrake*

the brake chambers to exhaust to atmosphere. The mechanical linkage is also returned to the 'off' position.

Hand control valve (TC2) (BW)

On a rigid vehicle, the TC2 hand control valve can provide the means of operating the secondary system, but with this type of control an orthodox handbrake is also necessary. On articulated vehicles, it can be fitted as a trailer brake only, or as a means of operating the trailer brakes and the front brakes of the tractor.

The hand control valve enables the driver to stop the vehicle by using the secondary braking system, if the service brake (footbrake) has failed.

As shown in Figure 6.84, when the hand lever is applied, the cam, followers and spring move the piston down, which closes the exhaust valve. The inlet valve is opened admitting air from the reservoir, into the line and to the brakes. When the air pressure below the piston rises sufficiently to cut off further air supply, the exhaust remains shut, to prevent loss. So, the brake valve is in a balanced position with equal pressure above and below the piston and a required amount of air held in the brake chambers. When the handle is moved to the 'off' position, the exhaust valve stays open, allowing the pressure to escape.

This valve should *not* be used for parking and is marked to this effect.

Spring-brake control valve (truck) (BW)

This is a hand control valve, but is not interchangeable with the normal hand control valve (TC2). It is able to provide what is termed *graduable inverse air pressure*, which in simple language means that it can let air out of a brake chamber, as well as put it in, in a controlled fashion.

As shown in Figure 6.85, when the hand lever is on the 'off' position, the cam follower (A) causes the cam (B) to compress spring (C), which holds down piston (D). This piston pushes down onto the inlet/exhaust valve (M), opening the inlet at (E) and closing the exhaust at (N), allowing air from the supply port (G) into chamber (J), then out of delivery port (H).

The air pressure delivered is governed by the spring loading of (C), which is preset with the adjusting ring (Q). The pressure delivered has to be sufficient to hold the spring, in the brake actuators, fully compressed during normal driving.

When the hand lever is moved towards the secondary/park position, cam (B) moves up and reduces the pressure on spring (C). Piston (D) follows, forced by air pressure and return spring (K). This closes the inlet valve at (E) and opens the exhaust valve (N) allowing air from delivery port (H) to flow through aperture (O) into chamber (P), out of exhaust port (F), escaping to atmosphere, and causing the spring brakes to apply. The resultant reduction in air pressure below piston (D), within the hand valve, continues until spring (C) moves the piston down, closing the exhaust, and preventing a further drop of pressure in the delivery line.

As forces above and below the piston are equal, the valve is in the lapped (or balanced) condition, with both

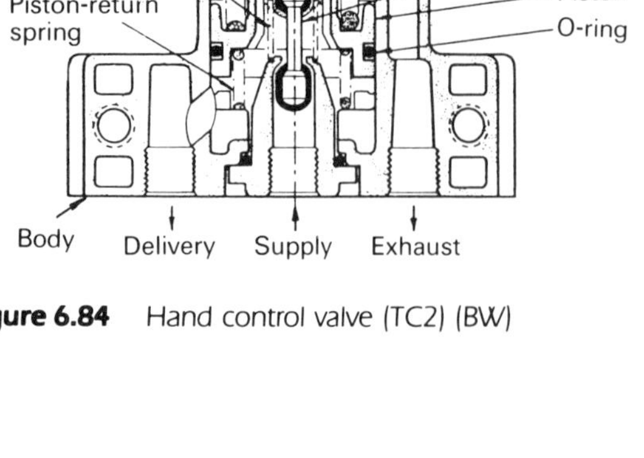

Figure 6.84 Hand control valve (TC2) (BW)

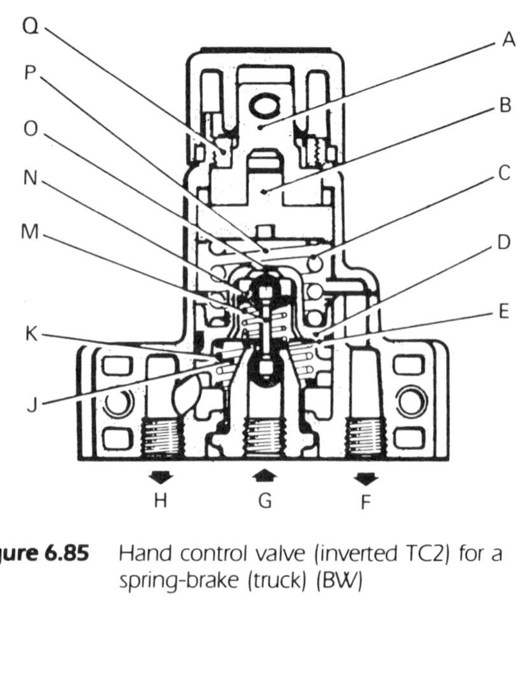

Figure 6.85 Hand control valve (inverted TC2) for a spring-brake (truck) (BW)

inlet and exhaust valves closed. This unit is self-lapping in all positions of the lever.

When the hand lever is moved further towards the secondary/park position, the above operation is repeated, until eventually all the air has been exhausted from the spring-brake chamber and the springs are fully extended, fully applying the brakes. There is a handle lock at the park position.

When the hand lever is moved towards the 'off' position, air pressure is gradually applied until the springs are held off.

Spring-brake hand control valve (L4) (BW)

This type of hand control valve can cater for a tractor with spring brakes, connected to a trailer with conventional brake chambers, without the use of an inverse spring-brake valve or similar device. Two relay valves are included in the piping system so that the hand valve only needs to supply a signal pressure to each relay valve, and this speeds up the brake application.

As shown in Figure 6.86, there is an L-shaped gate, within which the lever is moved. The horizontal line has the 'off' position on the right (when viewed from the front) and the 'secondary' position at the left-hand end, whilst 'park' is at the upper end of the vertical slot. A spring-loaded collar around the lever, has to be pulled against the spring, to enter or leave the park slot. (**Note:** An L-gate with 'off' at the left-hand end is also produced.)

As shown in Figure 6.87, the unit is fitted with two vertical, parallel, graduating elements. One of the elements controls upright air and the other controls

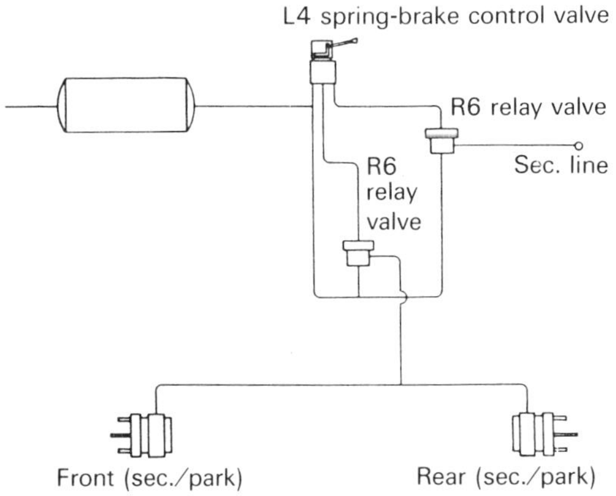

Figure 6.86 Typical piping layout for L4 hand control valve

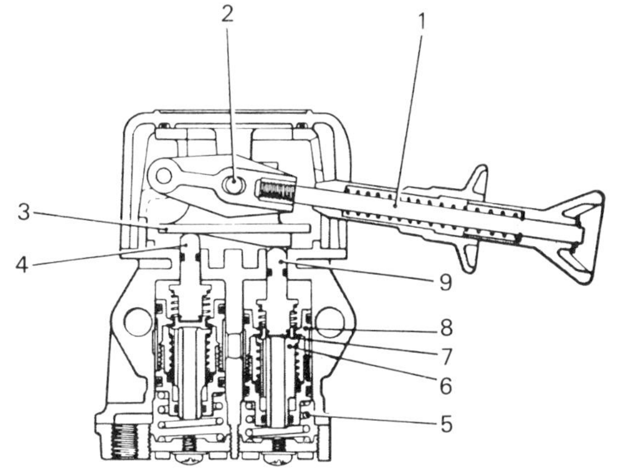

Figure 6.87 Hand control valve (BW)

inverse air. When the handle is in the 'off' position, the upright air valve delivers nothing to the brake actuators, whilst the inverse air valve delivers air to hold off the springs. When the lever is moved towards the secondary mark on the casting, this action rotates cam (3) which pushes down plunger (9) of the upright element, which closes the exhaust valve (7), opening inlet valve (6), so that compressed air is allowed to pass to the orthodox brake chambers. As the pressure in the brake chambers rises, the pressure, which also acts down upon the upper piston (8), causes it to move down, against the graduating spring (5) and so closes the inlet valve. When the lever is moved further towards the secondary marking, the air pressure is gradually increased, until at that position maximum pressure is achieved. Whilst this has been happening on the upright element, on the inverse side of the valve the reverse has been taking place, until at the secondary mark, zero pressure remains in the line to the spring chambers.

When the lever is moved from secondary to park position, the cam (3) is lifted up by the lever (1) and pivot (2), thus allowing upright plunger (9) and the inverse plunger (4) to rise and allow all air to exhaust from *both* spring chambers and from the conventional brake chambers.

Note: With this hand valve in the park position, only the spring brakes are applied. Thus, in the case of an articulated vehicle with spring brakes on the tractor and orthodox air brakes on the trailer, the combination is held at rest *solely* by the spring brakes on the tractor.

RM2 handbrake valve (WO)

This valve is used on tractor and trailer combinations, where spring brakes are fitted to the tractor (see Figure

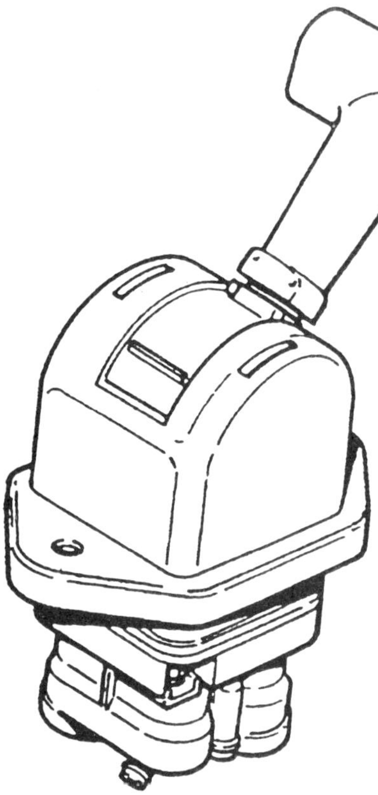

Figure 6.88 Pictorial view of RM2 handbrake valve (WO)

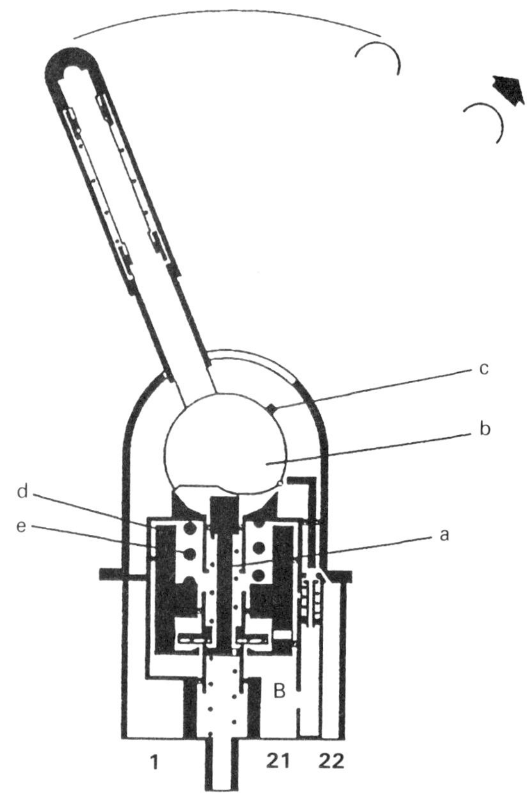

Figure 6.89 Section view of RM2 hand valve

6.88). The handle is locked in position, when park is selected, but when released it is automatically returned to the 'off' (driving) position, by the action of a powerful spring.

Driving

As shown in Figures 6.89 and 6.90, with the handle in the 'off' (drive) position, inlet valve (i) is open, allowing air from port (1) to flow to port (21) via chamber (A), and to port (22) via drilling (j).

Emergency

As shown in Figures 6.89 and 6.90, with the handle in the emergency (or secondary position), i.e. held by hand between the two extremes of the gate, cam (b), to which the handle is attached, depresses piston (a) and valve seat (g), closing inlet valve (i). Consequently, exhaust valve (h) is opened, which causes the reduction of air pressure below piston (d). When the downward force of spring (e) exceeds the upward force caused by air pressure in chamber (B), piston (d) moves down, together with valve seat (g), closing exhaust valve (h). This provides a lapped (balanced) condition, with the brakes applied in proportion to the amount that the hand lever is moved by the driver.

Figure 6.90 Hidden detail of RM2 hand valve

Note: Ports 21 and 22 are linked via chamber (C) and drilling (m).

Parking

With the handle in the park position, the handle is locked mechanically. Exhaust valve (h) is fully open, and port (1) is sealed from chamber (B). Peg (c), on cam (b), moves down piston (k), to close valve (l), closing routes to ports 21 and 22. Air pressure in chamber (B) now passes to port (22) via drillings (j) and (m). Consequently, the spring brakes on the tractor are fully applied, and in addition the trailer brakes are applied via port (43) on the trailer control valve (see Figure 6.91).

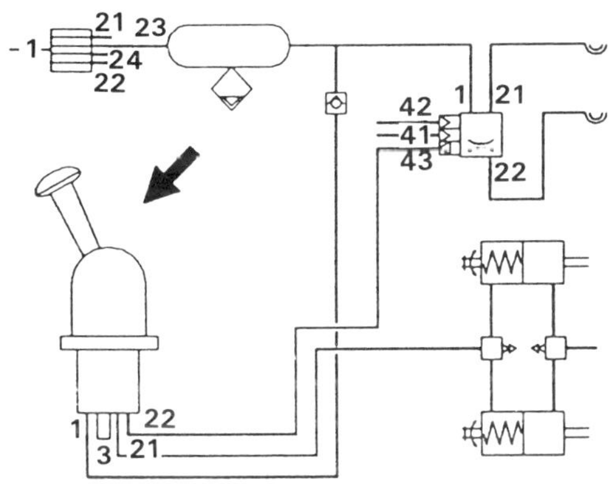

Figure 6.91 Piping diagram with trailer control valve

Differential-protection valve (BW)

This type of valve is fitted into a spring-brake system to prevent the compounding of forces and consequently the overloading of the spring-brake push-rod, slack adjuster and foundation brake, that would be caused by the severe application of the service brake, if the spring (secondary) brake was already in use or vice versa.

As shown in Figure 6.92, when the vehicle is in the driving condition, the hand control lever is in the 'off' position, so that compressed air is supplied to the spring-brake actuators to keep the spring compressed. This air passes through the differential-protection valve, entering at port (D) and leaving through ports (A) and (C) to reach the actuators, and hold-off the springs.

If the service brake is applied and held-on, then the hand lever moved to the secondary position, air will exhaust to atmosphere through the hand valve in the normal way. However, as the pressure in the secondary line falls, the service pressure becomes the higher of the two lines connected to the differential-protection

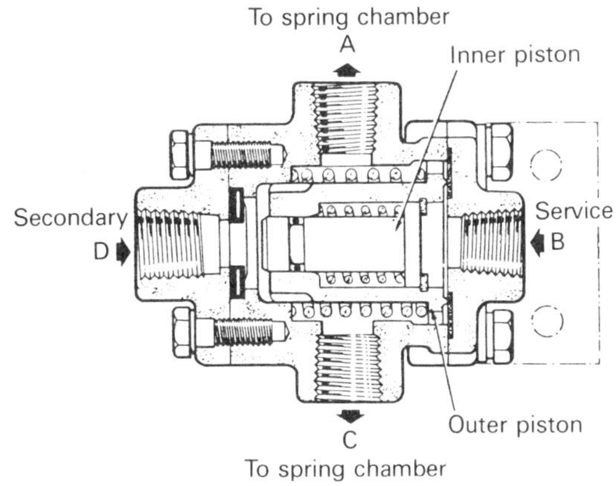

Figure 6.92 Differential-protection valve (BW)

valve, and so pushes the inner piston across to seal secondary port (D). The outer piston then unseats, allowing compressed air from the service line to pass through ports (A) and (C) and so reach the springs, preventing them from extending whilst the service brake is in use. When the pressure in the service line drops, first the outer piston and then the inner piston moves back under spring pressure, to seal off service port (B) and to open secondary port (D), enabling the spring chambers to exhaust. Thus, as the service brake is released, the secondary brake is applied simultaneously.

If the service brake is applied when the secondary brake is already applied, for example the secondary line is already devoid of pressure, the pressure in the service line moves the inner piston across to close off the secondary line. If this application exceeds the secondary pressure by approximately 30 lbf/in^2 (207 kN/m^2, 2 bar) (known as the *pressure differential*) the outer piston will be moved across, to allow air from the service line to enter the spring-brake chambers and hold off the springs, so that the secondary system is not applied.

When pressure is subsequently reduced in the service line, the outer and inner pistons are reseated by their springs, sealing off port (B), opening port (D) and allowing the air to exhaust from the spring chambers, so that the springs again apply the brakes.

Overload-protection valve (WO)

This is an anti-compound valve, used where there are spring brakes for secondary braking. It prevents the force of both the spring and service brakes being

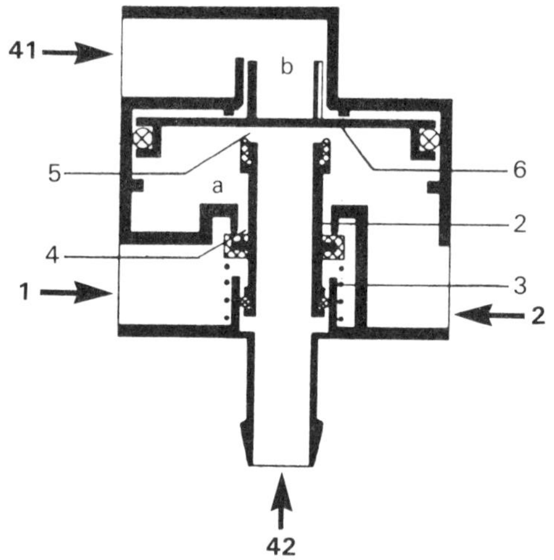

Figure 6.93 Overload-protection valve (WO)

applied simultaneously, thereby avoiding the otherwise inevitable damage.

As shown in Figure 6.93, port (41) is linked by a T-connection from the service line between the foot valve and the brake chambers. Port (1) is connected to a reservoir; port (42) is connected to the hand valve; and port (2) is connected to the spring-brake chambers.

All brakes 'off'

As shown in Figure 6.93, compressed air from the hand valve, enters through port (42) and passes through the open inlet/exhaust valve (5) into chamber (a) and on to the spring brake chambers via port (2). Port (41) and chamber (b) are at zero pressure. Inlet valve (4) is held shut by reservoir pressure and return spring (3). Piston (6) is held at TDC.

Service brake only applied

As shown in Figure 6.93, compressed air from the foot valve enters port (41) and chamber (b), but because there is already equal pressure below piston (6) (to hold the spring brakes 'off'), this piston will remain at TDC. Thus, the overload-protection valve is inoperative at this stage, although the service brake will operate normally and the spring brake remains in the 'off' position.

Spring brake only applied

As shown in Figure 6.93, as the foot valve is *not* being applied, port (41) is not receiving compressed air and therefore piston (6) remains at TDC. When the hand valve is operated, air from the spring brake flows back, via port (2), inlet/exhaust valve (5) and port (42), to exhaust from the hand valve exhaust port, thus allowing the spring to apply the foundation brake.

Simultaneous use of both service and spring brake

There are two ways in which this could happen:

a) Spring brakes are already applied and then the service brake is used;
b) Service brake is applied and then the spring brakes are used.

The anti-compound action taken by this valve, in each of the above cases, is explained below and with reference to Figure 6.93.

Spring brake applied and then service brake is used

As shown in the spring brake only applied section above, piston (6) is at TDC and chamber (a) is at zero pressure. When the foot valve is applied, compressed air enters port (41) and chamber (b) to act upon the upper surface of piston (6). Lacking pressure below, piston (6) is forced downwards by the air from the foot valve, closing inlet/exhaust valve (5) and opening inlet valve (4). Consequently, compressed air from the reservoir, enters port (1), passes through chamber (a) and port (2) en route to release the spring in the spring-brake chamber. Although neither the driver nor the fitter can see this occur, the spring brake has been released and the service brake left-on. When there is only a light service-brake application, the spring brake is only partly released. The overload-protection valve is in the lapped or balanced condition, as soon as the pressure in chamber (a) exceeds the pressure in chamber (b), i.e. piston (6) moves up and inlet valve (4) closes.

Service brake applied and then the spring brakes are used

As shown in the service brake only applied section above, piston (6) is at TDC. However, if the spring brake is then applied, pressure will drop in chamber (a), beneath piston (6). That piston will descend, owing to the pressure in chamber (b) above, from the service line, causing piston (6) to close the inlet/exhaust valve

(5) and open inlet valve (4), to allow compressed air to flow to the spring brakes, via port (2).

Naturally, there are various degrees of brake application and with a light application of the spring brakes, only a small amount of compressed air is necessary to restore the balanced (lapped) condition, i.e. where there is equal pressure above and below piston (6).

If the driver releases the service brake (leaving the secondary control applied), pressure above the piston drops and the pressure retained below causes the piston to rise. This action opens the inlet/exhaust valve (5), providing an exhaust route from the spring brake to the hand valve exhaust port, via ports (42) and (2). This will leave the spring brakes applied to whatever degree the hand control valve was originally set.

Double-check valve (CD)

The double-check valve, or shuttle valve, is incorporated in the line, between the control valves and the brake chambers. When only one control valve is operated, the check valve blanks off the line connected to the other control valve and prevents air from being exhausted through this valve.

As shown in Figure 6.94, when compressed air enters at port (A) of the check valve, it causes the shuttle to move to the opposite end of the valve, sealing port (B), but permitting the air to leave by port (C). The reverse action would take place if air pressure was introduced to port (B), i.e. port (A) would be sealed and the air would leave by port (C). However, if both ports (A) and (B) are pressurised the priority would go the highest pressure, which would seal off the opposite inlet port and yet still use the same port (C) for delivering this higher pressure to the brake chambers.

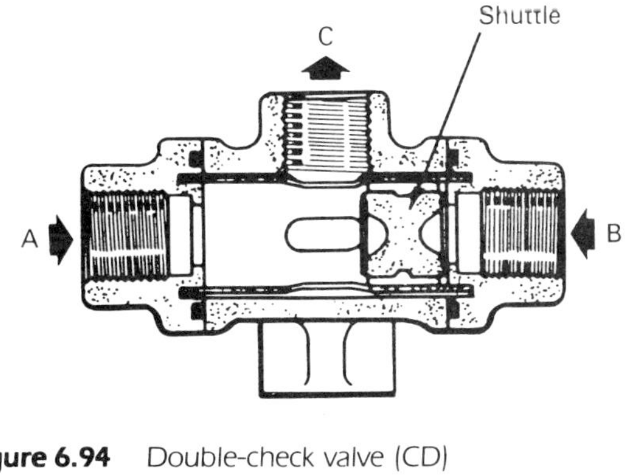

Figure 6.94 Double-check valve (CD)

Biased double-check valve (WO)

Although this valve serves the same purpose as an ordinary double-check valve, by allowing two independant sources of supply to charge a single control line, this unit proportionately reduces the air pressure that is delivered by one supply source.

As shown in Figure 6.95, when air is supplied to port (M1), graduating piston (1) will be forced against the stop at the opposite end of the valve, and air will pass out of port (S). Input and output pressures will be identical.

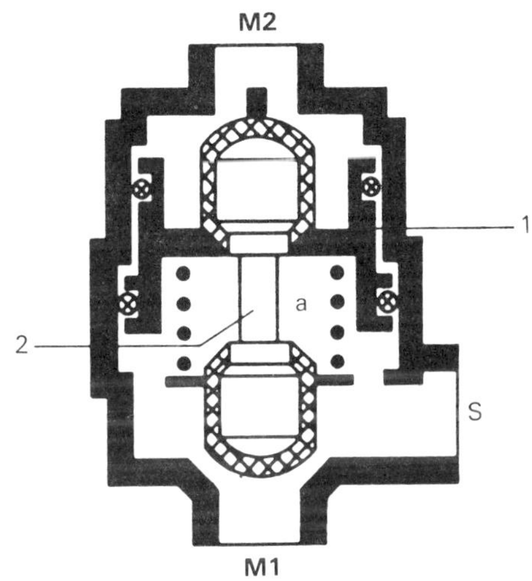

Figure 6.95 Biased double-check valve

When air is supplied to port (M2), graduating piston (1) will be moved down. The bottom section dumb-bell valve (2) will seal port (M1) whilst its top section will uncover port (M2), allowing air to enter chamber (a) and leave via port (S).

It should be noted that the upper and lower face areas of the graduating piston are not identical. The lower face has an area 1.3 times greater than the upper face.

Remembering that Force = Pressure × Area, it can be seen that a low air pressure acting upon the bottom face of the piston will cause movement, whereas a comparatively high air pressure is necessary, at the top face, to cause movement. Consequently, as air pressure from port (M2) is passing through the valve, to leave by port (S), it is also acting upwards beneath the lower face of the graduating piston. This will cause the piston to rise, shutting off the inlet port, but leaving the pressure within the valve, at port (S), and beyond. The valve will be in the lapped (or balanced) condition and will remain so until the control is released.

The R6 relay valve (BW)

The purpose of a relay valve is to speed up the application and release of actuators, by bringing reservoir pressure close to the particular actuator, and then activating the relay by a signal pressure from a control valve. The signal is fast-acting because only a small volume of air is required. A common application is its use, on a long vehicle, to operate the rear brakes.

As shown in Figure 6.96, this unit consists of a base, containing an inlet/exhaust valve and ports for connections to reservoir, actuators and exhaust. The upper half, forming a cover, has a port or a connection to the control valve. This upper section also contains a close-fitting piston and its return spring.

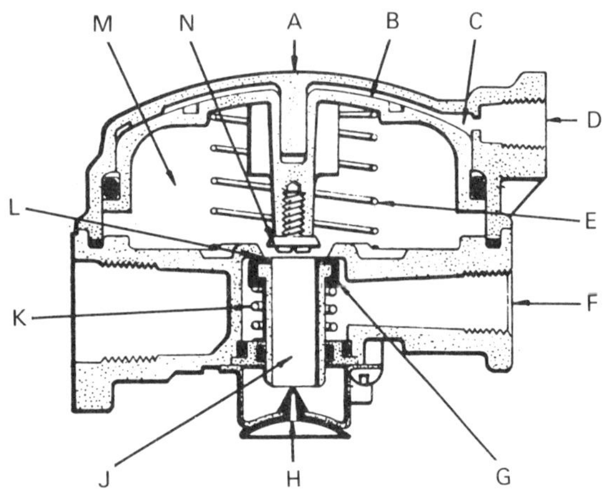

Figure 6.96 Relay valve (R6)

When the uppermost port (D) receives a signal pressure from the control valve, this compressed air enters this valve between the top casting (A) and the top of the piston (B) and as a result the piston moves downwards (only a small amount of pressure is necessary to cause this large diameter piston to react). The downward movement of the piston brings exhaust valve seat (N) into contact with the inlet/exhaust valve (G), and seals off the exhaust passage (J). The inlet/exhaust valve is moved downwards, against return spring (K), opening the inlet valve (L), so that air from the reservoir port (F) can enter the valve, passing through the chamber (M) and leaving by one of the delivery ports en route for an actuator. This flow of air continues until the combination of forces of both return springs and the air pressure below the piston equals the force of the air pressure above the piston. At this stage the piston will rise sufficiently to close the inlet at (L). The valve is now in the lapped condition with both the inlet and exhaust valves closed.

If the driver reduces the amount of movement that has been made on the control valve, the signal at (D) will decrease and the forces below the piston will cause it to rise until the valve seat (N) is lifted above the valve and air will escape to atmosphere through exhaust passage (J) and past rubber flap (H). The process continues until the forces above the piston are equal to those below the piston and the exhaust valve has closed, when again the valve is said to be in the lapped condition.

Relay valve and three-way, double-check valve (R62DC) (BW)

As shown in Figure 6.97, to avoid overcrowding this figure, the service signal port (nearest to the reader) has been omitted.

The R62DC relay valve has been developed by bolting a three-way, double-check valve to the top of an R6 relay valve, and by making a drilling in both the double-check valve and the relay valve, to provide connecting airway (C) between the two valves.

This type of relay valve can deliver compressed air to the control line of a trailer, in response to a signal from any one of three sources, whereas a standard R6 relay valve can only receive signals from *one* source. A typical arrangement for the R62DC relay valve is to have a signal connection from each half of a dual foot valve and another from the hand valve.

Double-check valve (E) contains two shuttles, (F) and (H) which work within guides (G) and (J). Drilling (C) provides a connection between the double-check valve and the relay valve, at a point above relay piston (L).

Note: Guides (G) and (J) are slotted to allow air to flow to drilling (C).

When the foot valve is applied, signal pressure enters both ends of the cross leg of the R62DC, and acts upon shuttle (F). If the pressure is equal from both sources, shuttle (F) will finish at a central position, but if one signal is stronger than the other, the shuttle will finish at the end producing the least pressure. Whatever the result, compressed air will then pass through the slots in guide (J), into the bore of the other leg, and move shuttle (F) to the end of guide (G) that is nearest to the secondary signal port, to seal off that port.

The compressed air, from the foot valve signal, then

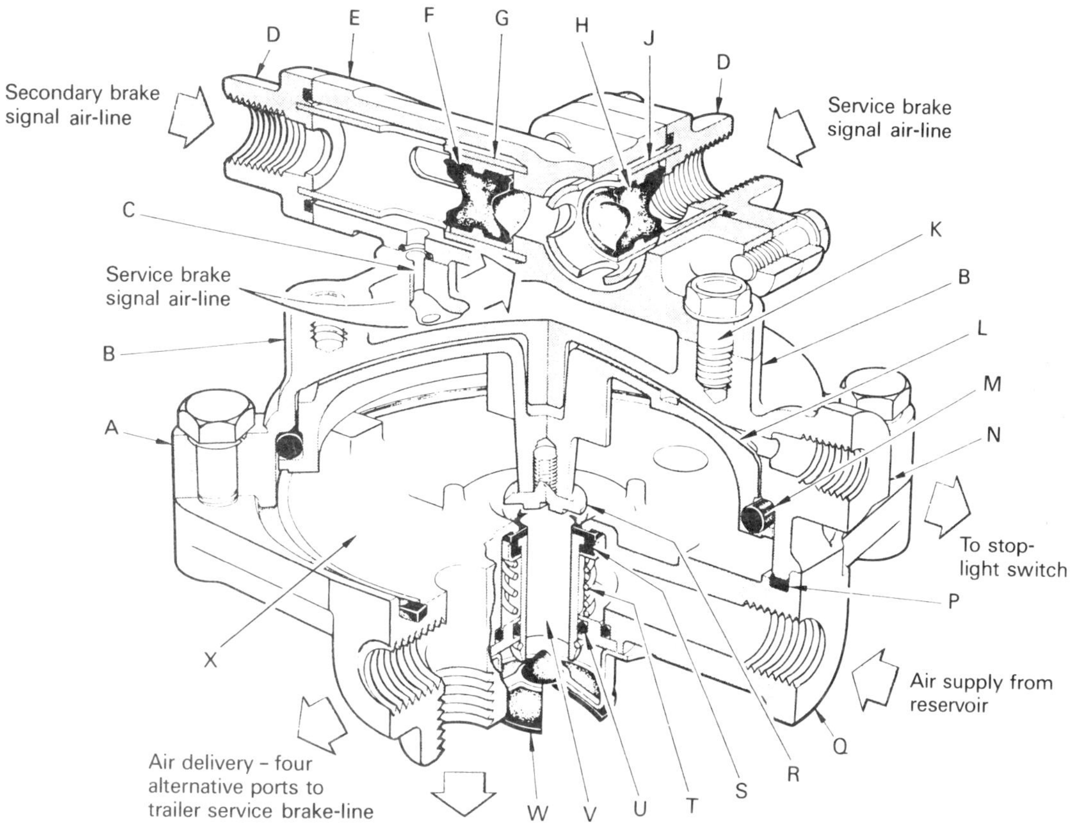

Figure 6.97 R62DC relay valve (BW)

passes through drilling (C), to act upon and move down relay piston (L). From this point the relay valve acts exactly the same as a conventional R6 relay valve.

When a secondary brake application is made, the signal pressure acts upon shuttle (F) and causes it to move to the extremity of guide (G) to seal off the service bore. Compressed air passes through drilling (C) to act upon piston (L), which then acts in the same manner as in the conventional R6 relay valve.

A variation of the R62DC relay valve, is the R61DC (see Figure 6.98), which only has a two-port, double-check valve attached. The operating principles are the same as previously described, but this valve is only able to receive signals from two different sources.

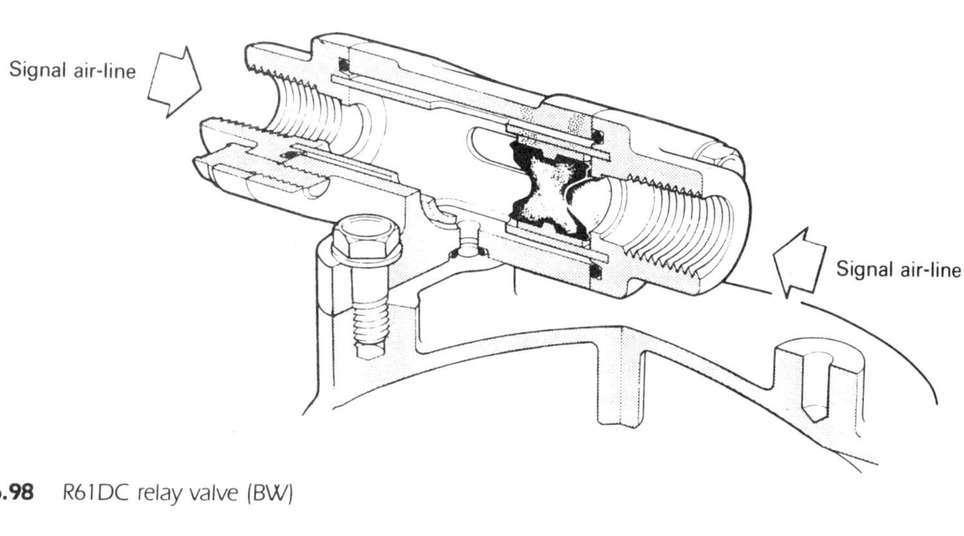

Figure 6.98 R61DC relay valve (BW)

Multi-relay valves

Dual and triple-relay valves can apply the brakes by directing air from the reservoir to the brake chambers, without having to pass through either the foot or hand valve. The principle is the same as the single-relay valve, i.e. that reservoir pressure is connected to the lower port of the relay valve and a signal pressure is sent from a control valve (foot or hand) to an upper port, when the brakes are to be applied. In the case of multi-relay valves the signal pressure required to trigger the valve can be supplied by two or three sources, although the failure of one of these will not affect the other systems, even though they are all connected to the same relay valve. (A dual valve can be used to control three lines by the use of a double-check valve, or change-over valve in the line.)

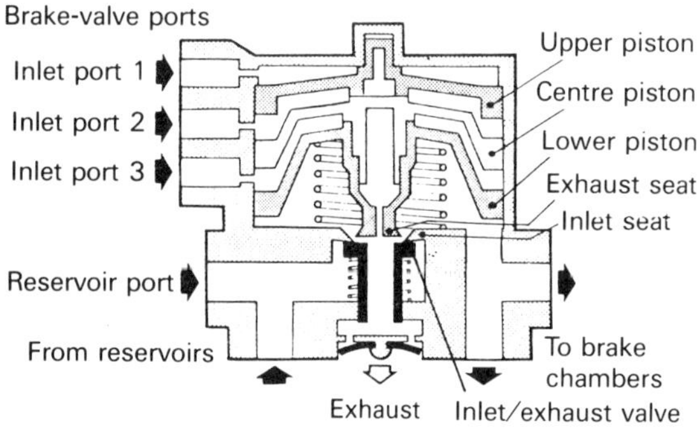

Figure 6.99 Triple-relay valve

As shown in Figure 6.99, if more than one control valve is operated simultaneously, the maximum pressure will depress the piston(s) within the relay valve, to release the air at the reservoir pressure, that is connected to the lowest port. When signal pressure from a control valve is received above one of the upper pistons, the piston moves downwards, closing the exhaust seat on the face of the inlet/exhaust valve and unseating it from the inlet seat, so that reservoir pressure is supplied to the brake actuator.

The compressed air passing to the actuator also acts upon the underside of the lowest piston, so that it moves up against the spring when the pressure becomes higher than the downward pressure. This is known as the balanced or lapped condition, when the pressure of the air delivered is equal to the pressure delivered by the driver's force onto the control valve. If additional pressure is applied, or pressure is released by the driver, this will be acted upon by the lower piston to maintain the correct requirements at the actuators.

When the control valve is released, all air above the relay piston is released from the exhaust of the driver's control valve, whilst all air pressure from the brake actuators is released from the exhaust at the base of the relay valve.

Adapter valve (WO)

The function of this valve, which is sometimes known as a *regulator valve*, is to reduce the braking force on a particular axle during a partial brake application, and to act as a quick-release valve when necessary.

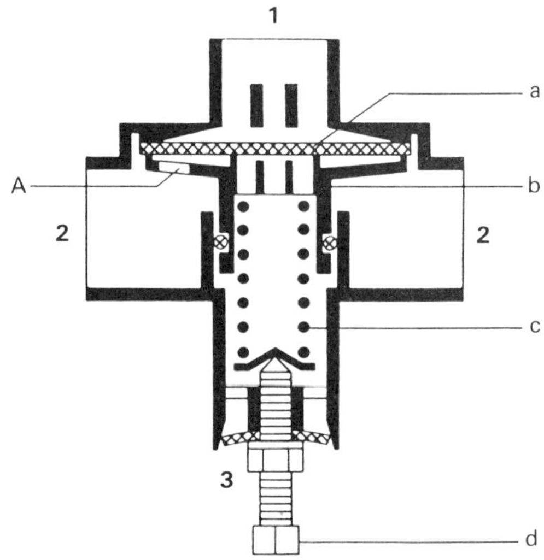

Figure 6.100 Adapter valve (WO)

As shown in Figure 6.100, when the brakes are not in use, spring (c) keeps piston (b) at TDC, where diaphragm (a) seals off the route between port (1) and ports (2).

When the brakes are applied, compressed air enters port (1) and builds up against the upper face of diaphragm (a), causing the diaphragm to exert a greater downward force than the upward force exerted by spring (c) below. Piston (b) moves down and allows compressed air to pass over the edges of diaphragm (a), into ports (2) en route to the brake chambers.

As the air pressure rises in ports (2), it assists spring (c) by acting beneath diaphragm (a), until the pressure/spring force equals the pressure above, when the piston and diaphragm are returned to the top of their stroke. A balanced position is thus achieved, with the brakes applied and held in that condition.

When the air pressure is increased at port (1), the

piston (b) and diaphragm (a) are again moved down against force of spring (c) enabling the incoming air to reach the brake chambers without sustaining any pressure loss.

When the pressure from the control valve reduces, the returning pressure from the brake chambers lifts diaphragm (a), allowing this air to pass through aperture (A), to reach the hollow piston and pass to atmosphere via exhaust port (3).

The pressure necessary to open this valve can be adjusted by setscrew (d).

Regulating valve (WO)

For light brake applications, this valve will reduce the air pressure, and consequently the braking force, at the regulated axle. The need for such control is often where draw-bar trailers are continually operating in hilly country, with long, downhill sections. In such conditions, the trailer front-axle brakes will become hotter and there will be greater lining wear on light brake application than occurs at the rear-axle brakes. Full brake applications will still provide maximum design braking force to the brakes of both axles. It also acts as a quick-release valve.

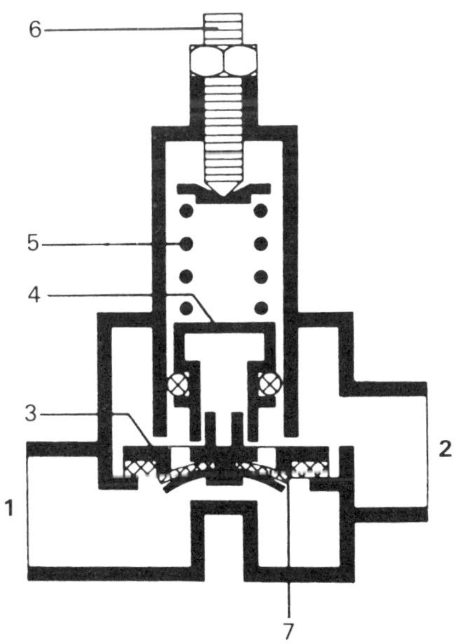

Figure 6.101 Regulating valve (WO)

As shown in Figure 6.101, valve assembly (3) is restrained on its seat by spring (5), acting against piston (4). This valve has an opening pressure of 6–30 lb/in^2 (0.4–2.0 bar; 40–207 kN/m^2), which can be regulated by adjusting screw (6). Compressed air enters port (1) and acts upon valve assembly (3), overcoming spring (5) and thereby causing the valve to open, allowing compressed air to travel to the brake chambers, via port (2). Simultaneously, air pressure acts upon piston (4). As this pressure increases, spring (5) is correspondingly compressed, until valve assembly (3) allows compressed air to flow to the brake chambers. When the brakes are released, air exhausts via check valve (7).

Quick-release valve (BW)

The rapid application and the rapid release of brakes is almost as important as the actual efficiency of the brakes themselves.

On long wheelbase vehicles and on trailers, the quick-release valve is a valve that can be fitted into the system close to the brake actuators, so that exhausted air does not have to travel all the way back to the foot or hand valve, merely to escape to atmosphere. In the case of spring-brake actuators, the quick release of air can aid the rapid application of the secondary brakes.

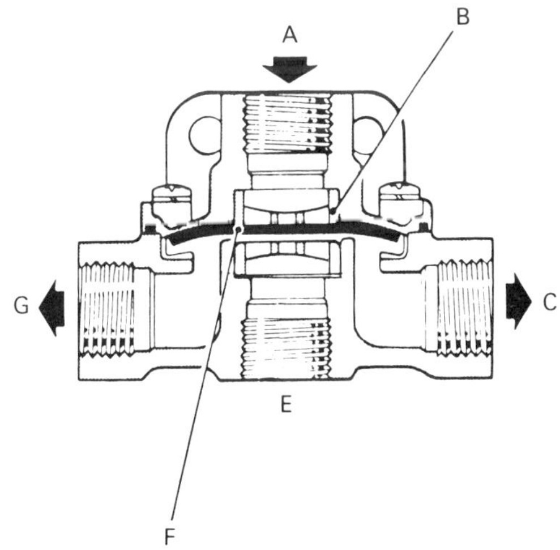

Figure 6.102 Quick-release valve – brakes applied (BW)

As shown in Figure 6.102, the construction of this valve is extremely simple, consisting of a diaphragm that is held between the two halves of the body. When the brakes are applied, air from the control valve enters port (A) and the diaphragm is pushed off its seat (B) and onto exhaust seat (F), thus preventing losses through the exhaust port (E). The compressed air passes out of ports (C) and (G) en route to the brake chambers.

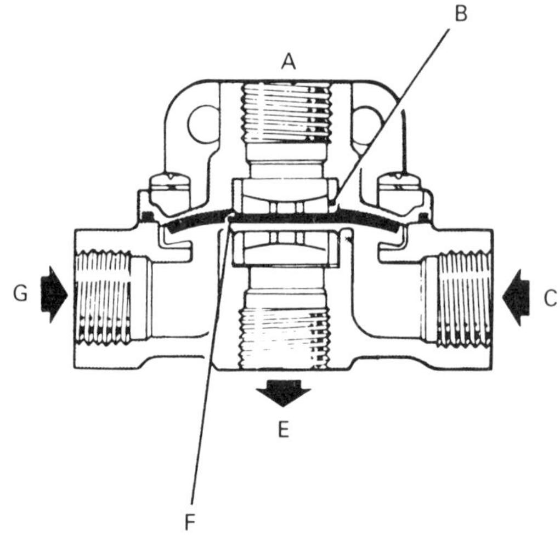

Figure 6.103 Quick-release valve – brakes released

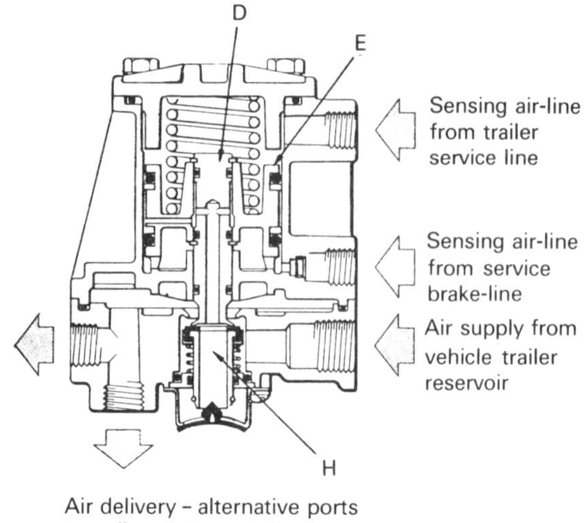

Figure 6.104 Supply-dump valve (BW)

As shown in Figure 6.103, when the pressure at (A) is reduced, the pressure within the actuator, and consequently also under the diaphragm, will be greater than that at (A). As a result, the diaphragm lifts off seat (F), allowing the air to exhaust to atmosphere through exhaust port (E).

Supply-dump valve (BW)

The EEC Directives stipulate that if the trailer service brake should fail, the next full application of the service brake will, within 2 seconds, automatically reduce the pressure in the emergency line to 22 lbf/in^2 (152 kN/m^2, 1.5 bar) which will cause the relay emergency valve to apply the trailer brakes automatically.

The role of this valve is to monitor the pressure in the service brake system of the towing vehicle and the pressure in the service brake line of the trailer. Should the pressure in the service line that serves the towing vehicle be considerably higher than the trailer service line, the valve will exhaust the trailer emergency line and this will activate the relay emergency valve to apply the trailer brakes automatically.

The construction is based upon the familiar principle of inlet/exhaust valve and a hollow valve stem, as shown on page 14. As shown in Figure 6.104, the valve body casting contains a piston (E), which is spring-loaded from above. Secured in this piston is a valve stem (D), which acts upon the spring-loaded inlet/exhaust valve (H).

Normal driving

As shown in Figure 6.105, as neither of the service brake systems are in use, no air pressure is entering this unit from either sensing line. The piston spring is set to give a downwards thrust equal to a pressure of approximately 20 lbf/in^2 (138 kN/m^2, 1.4 bar). The valve stem will depress the inlet/exhaust valve, so that the free flow of air from the vehicle trailer control reservoir is not obstructed, and the exhaust port of this valve remains closed.

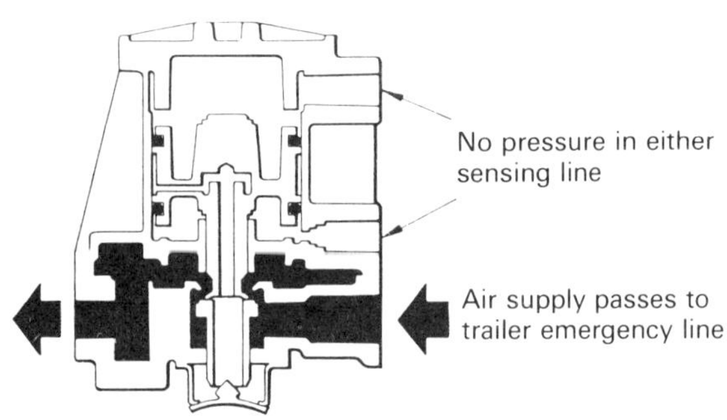

Figure 6.105 Supply-dump valve – brakes not applied

Service brake applied (no faults in the system)

As shown in Figure 6.106, equal air pressure from both sensing lines will act both above and below the piston. Therefore, it will not move from its original position. However, some initial movement is possible, due to momentary differences in pressure between these two lines, but the effect is limited by a choke in the sensing line from the service vehicle. The design ensures that this type of difference in pressure will never cause a full emergency brake application.

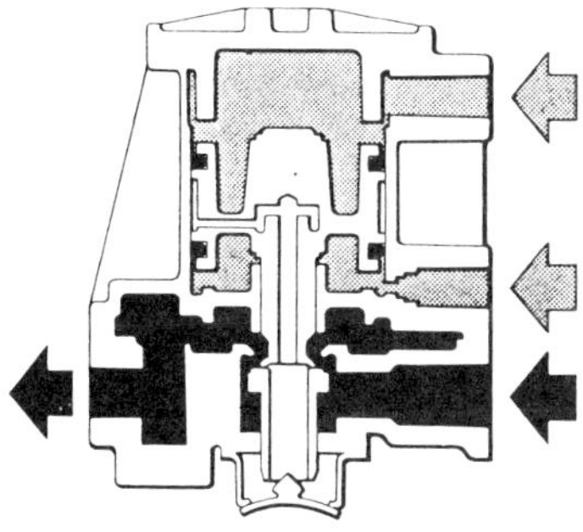

Figure 6.106 Supply-dump valve – service brake applied

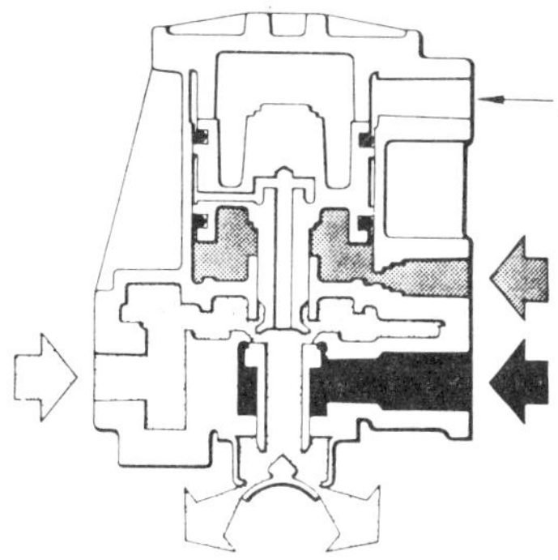

Figure 6.107 Supply-dump valve – failure of service-line pressure

Failure of trailer service brake line

As shown in Figure 6.107, the next full service-brake application will supply its sensing line with the same pressure, but the defective trailer line will not be able to match this pressure. Consequently, the pressure will rise, as will the valve stem attached to it, allowing the spring-loaded inlet/exhaust valve to seat, sealing off the supply to the emergency port of the unit. As the piston valve stem continues to rise, it uncovers the exhaust port so that air in the pipeline between this unit and the relay emergency valve can exhaust to atmosphere via this hollow stem. The resulting pressure drop in the emergency line to the relay emergency valve causes this valve to apply the trailer brakes automatically using air pressure from the trailer reservoir.

Caution: When the vehicle and trailer are braked in this fashion, great care should be taken when correcting the situation.

Once the service brake pressure is released, this valve is restored to the normal running condition as the relay emergency valve is again supplied with a hold-off pressure. Thus, unless a handbrake or secondary brake is applied, this action provides a potential run-away situation.

It is of the utmost importance that the trailer service line should be repaired before driving the combination any distance. The use of the combination in this defective condition is dangerous, because even a moderate application of the service brake will cause full trailer braking, and frequent braking would produce a situation where the trailer reservoir was exhausted, leaving the drawing vehicle to provide all the braking.

The piping layout for a supply-dump valve is shown in Figure 6.108.

Trailer control valve (WO)

This type of valve is fitted to towing vehicles, to act as a

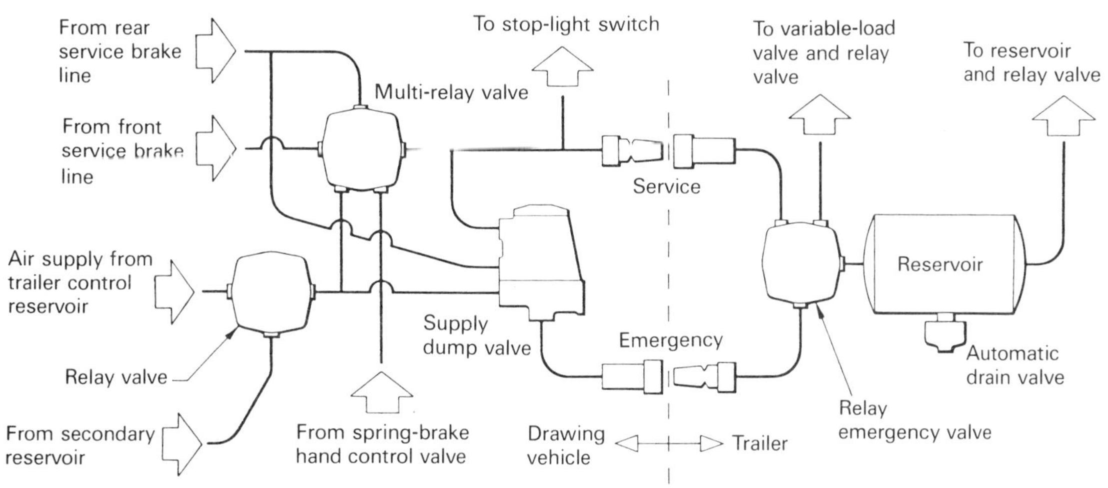

Figure 6.108 Piping layout for supply-dump valve

multi-relay valve, supplying compressed air, as necessary, to both the yellow (control) and red (supply) coupling heads of a two-line braking system for a trailer.

There are many of variations of this valve, but the common types are discussed below.

First, a basic type which caters for both signal lines from a dual foot valve, to allow air pressure to be passed to the trailer control (yellow) line. In the event of one foot valve circuit failing to operate, it ensures that the remaining sound circuit will operate the trailer brakes.

Often there is an additional facility to allow the secondary hand control to operate also the trailer brakes, via the trailer control valve and the control (yellow) line.

A differential device can be included to ensure that the air pressure being passed to the trailer brakes is regulated by the valve that senses the weight of the load being carried by the drawing vehicle.

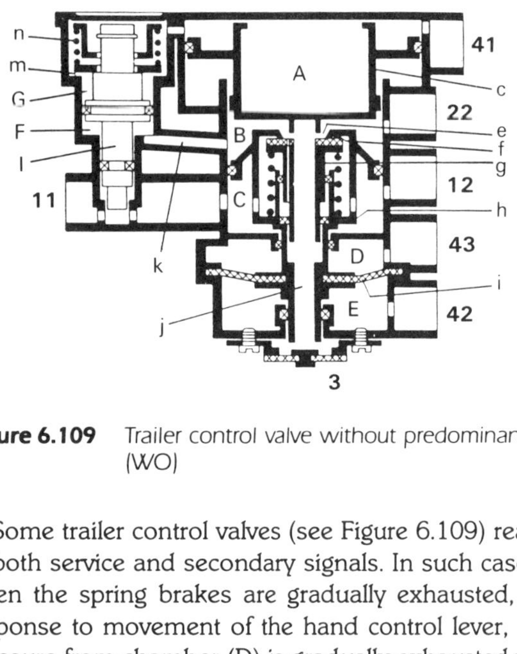

Figure 6.109 Trailer control valve without predominance (WO)

Some trailer control valves (see Figure 6.109) react to both service and secondary signals. In such cases, when the spring brakes are gradually exhausted, in response to movement of the hand control lever, air pressure from chamber (D) is gradually exhausted via port (43). As the supply pressure is then greater in chamber (C), it forces piston (h) upwards, closing exhaust (e) and opening inlet (f), so allowing air pressure to enter the control (yellow) line to operate the trailer brakes.

Most trailer control valves incorporate a supply dump arrangement, to ensure that should the pressure fail in the control (yellow) line to the trailer, that at the next full application of the footbrake treadle, the pressure in the supply (red) line would be reduced, thereby activating the relay emergency valve on the trailer, to cause the trailer brakes to be fully applied.

Another arrangement is to give *predominance* to the trailer brakes, by supplying air to the trailer brakes at a higher pressure than that supplied to the brakes on the drawing unit. Although some versions provide a predominance of 1 bar, there are other models where the increase is merely a fraction of this figure.

Figure 6.109 shows a basic model with a 2+/1− characteristic, i.e. it caters for both circuits of a foot valve (using upright air) and is also activated by the spring-brake hand control fitted to that towing vehicle (using inverse air).

Compressed air from a storage reservoir enters port (11), and acts on the underside of piston (l) causing it to rise to its upper limit, against the restraint of return spring (n). This air then passes via chamber (C) and port (12), to the supply coupling head (red) line, en route to the storage reservoir(s) on the trailer.

When the foot valve treadle is applied, air from its No. 1 signal circuit enters the trailer control valve at port (41), and fills chambers (A) and (G), and then acts on pistons (c), (d) and (l), causing pistons (c) and (d) to both move downwards. Piston (c) contacts valve (g), closing exhaust (e) and opening inlet (f). Supply air in chamber (C) then enters chamber (B) and then leaves via port (22) to enter the trailer control coupling head and the control (yellow) line, en route to the trailer brakes.

Simultaneously, air enters passage (k) and passes into chamber (F), to act beneath piston (l). When the control pressure above piston (l) reaches approximately 58 lbf/in^2 (400 kN/m^2, 4 bar), the downward force is greater than the force below, and consequently the piston moves down onto shoulder (m). This process ensures that piston (l) moves regularly and avoids the possibility of seizure.

Figure 6.110 shows a model 2+ and predominance, i.e. it caters for both circuits of a foot valve, but not the hand valve, although it has a predominance feature.

In the type of the trailer control valve with predominance, the outlet pressure at port (22) is dependant upon the amount of preload of spring (b), so that the pressure of the air leaving port (22) can exceed that of the air entering port (41).

Pressure builds up in chamber (B) and acts beneath pistons (c) and (d). In the predominance version (Figure 6.110), piston (c) moves up against air pressure in chamber (A) and spring (b), because of the difference in the sizes of pistons (c) and (d).

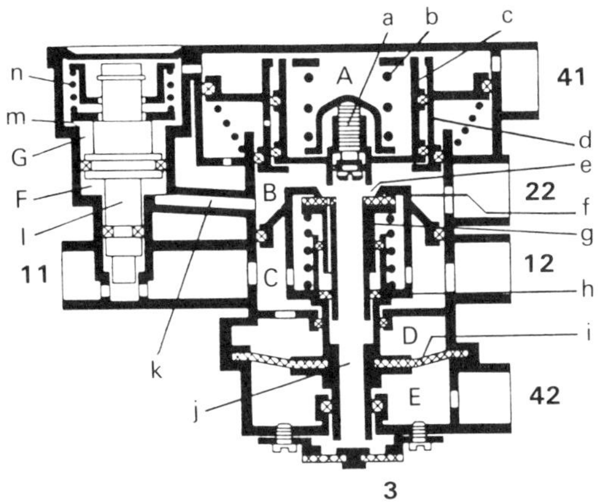

Figure 6.110 Trailer control valve with predominance

Note: The formula, *Force* = *Pressure* × *Area*, is often used in the design of valves. Consequently, valve (g) closes inlet (f), thus producing a balanced condition.

With a *full* application of the brakes, the control pressure above piston (c) *predominates* and inlet (f) remains open. When signal pressure from foot valve circuit (1) enters port (41), signal pressure from foot valve circuit (2) enters port (42) and chamber (E) to act below diaphragm (i). As air pressure in chambers (B) and (D), above piston (h) and diaphragm (i), is greater than that below, the position of piston (h) is unchanged.

If footbrake circuit (1) fails, footbrake circuit (2) will continue to signal port (42). This pressure will enter chamber (E), beneath diaphragm (i) and move piston (h) and valve (g) upwards. Piston (c) will seal outlet (e) and open inlet valve (f), to supply compressed air via port (22) to the trailer (yellow) control line, to apply the trailer brakes. In these circumstances predominance is *not* provided.

Should the trailer control (yellow) line break between port (22) of the trailer control valve and the trailer, pressure will not build up in chambers (B) and (F). Piston (l) would be then forced down by the control pressure in chamber (G), severely reducing the flow of air from port (11) to port (12). A pressure drop in the trailer (red) supply line would be caused, when the footbrake treadle was operated, as the opening of inlet (f) by piston (c), would allow air to feed back into chamber (B), where the supply pressure would be then at a very low figure, because of restriction of port (11) by piston (l). This pressure drop in the supply (red) line would cause the relay emergency valve on the trailer to route full trailer reservoir pressure to the brake chambers on the trailer and thus apply those brakes automatically.

Load-sensing valves

These are mostly automatic valves, which are self-regulating according to the weight of the load carried on a vehicle or trailer, to which the particular valve is fitted.

Such valves are marketed under a variety of names, including variable-load valve, light-laden valve, automatic variable-load valve, load-sensing valve, empty-load valve, load-apportioning valve and manual light-laden valve. Some of these valves are incorporated into relay and relay emergency valves.

One variation is the slave variable-load valve, which is produced for use in split braking systems in conjunction with an orthodox variable-load valve. This avoids the necessity of fitting two variable-load valves and their respective linkage.

Variable-load valve (BW)

The purpose of the variable-load valve is to regulate the amount of braking applied to the axle, to which this unit is connected, so that braking is proportional to the load carried on that axle. Therefore, a fully-laden vehicle would be allocated maximum braking and an unladen vehicle would only receive limited braking on that axle.

This valve achieves its function by means of a flexible diaphragm, which has its effective area varied by the vertical position of radial fins that are attached to the body. The differing area produced varies the delivery pressures to the brake chambers.

As shown in Figure 6.111, the unit is mounted on the chassis, and operating lever (6) is connected by linkage to the axle. When the vehicle is unladen the road springs are barely compressed. Consequently, both the axle and the valve arm are in their lowest position. When the vehicle is laden, the road springs compress, so that the distance between the axle and chassis decreases, causing the lever to move upwards, making ball pivot (31) lift valve stem (20) and support the valve (16).

When the foot valve is applied, compressed air is supplied to the supply port, where it passes to chamber (15) and forces piston (3) downwards. Simultaneously, air passes from (15) through balance pipe (19), to enter the chamber beneath piston (36) and move that piston upwards to clamp ball pivot (31) against stem (20). By trapping the ball pivot, unnecessary wear of the internal parts of the valve is avoided, as these parts now will not move merely because of the action of the suspension, but only in response to the brake application. Because inlet valve (16) is supported on stem (20), it is lifted off

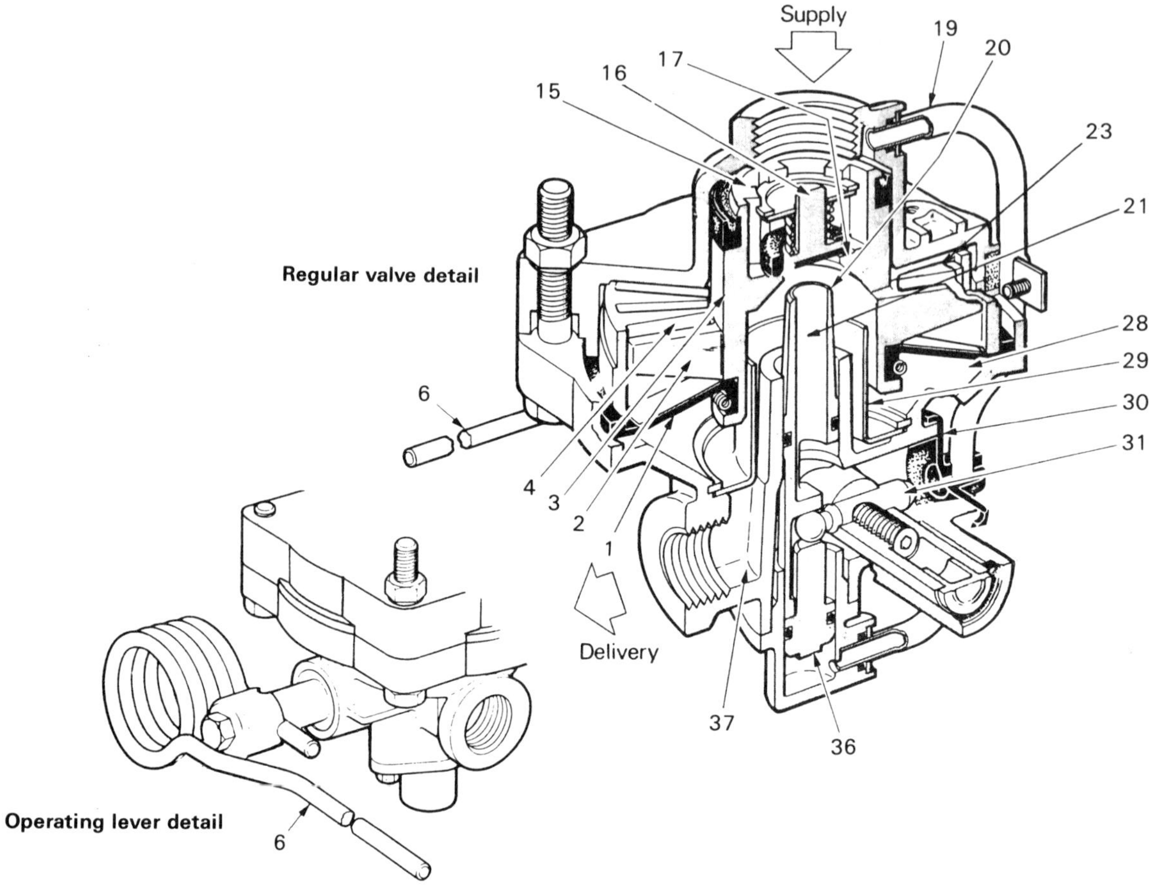

Figure 6.111 Variable-load valve (BW)

its seat (17) by the piston movement. This allows air to pass through delivery port (37), en route to the brake chambers.

Simultaneously, compressed air passes through the clearance between piston (3) and guide (29) into chamber (28), beneath diaphragm (1). As piston (3) moves down, diaphragm (1) starts to move away from the static support fins (4) and is partly supported by the fins (2), which are attached to the piston. Thus, the effective area of the diaphragm, supported by the moving piston is varied until the force under the diaphragm equals the force acting downwards above the diaphragm, when valve (16) will close. Therefore, balance is achieved, and yet the delivery pressure can be lower than the inlet pressure, owing to the control exerted over the effective area of the diaphragm, which initially is affected by the position of ball pivot (31).

When the control valve is released, so that the brakes are 'off', the pressure in the actuator line will cause the piston (3) to lift, so lifting valve (16) from stem (20) and allowing the air below the diaphragm to pass through the hollow stem, then past exhaust flap (30) and out to atmosphere.

Inshot variable-load valve (BW)

An inshot variable-load valve ensures that when the service brake treadle of an unladen vehicle is only lightly applied, sufficient air pressure is delivered to overcome the initial stiffness of the foundation brakes, and so avoid glazing the brake lining material. The term 'inshot' means supplying a 'shot' of compressed air. This is achieved by matching the pressure output of the variable-load valve to the pressure output of the foot valve, up to a maximum preset figure of around 10 lbf/in^2 (70 kN/m^2, 0.7 bar). Thus, the variable-load valve does not operate normally, until the foot valve is delivering in excess of 10 lbf/in^2 (70 kN/m^2, 0.7 bar). Above this figure, the pressure delivered to brake chambers, by the variable-load valve is relative to the load on the controlled axle of that vehicle.

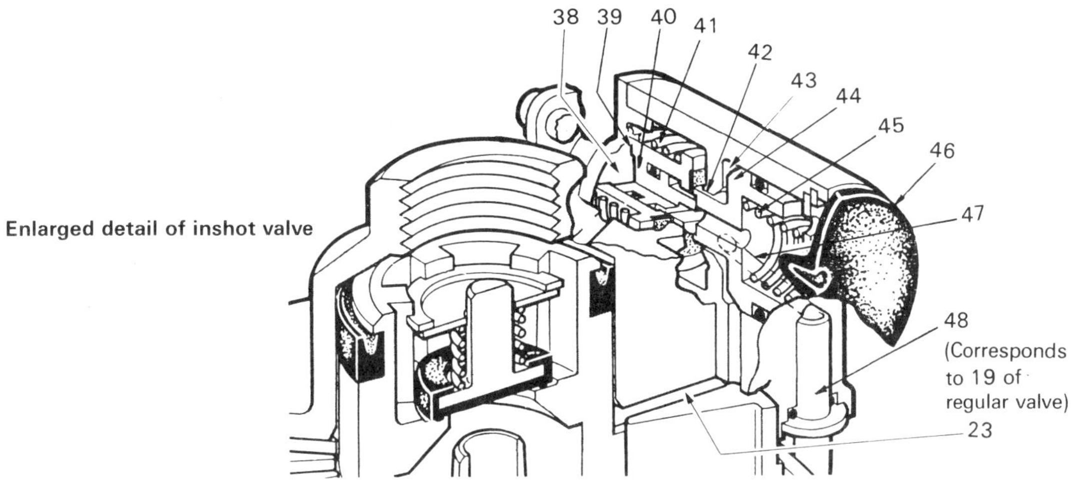

Figure 6.112 Inshot variable-load valve (BW)

Figure 6.112 shows the additional features added to a regular variable-load valve to convert it into an inshot version. See both Figures 6.111 and 6.112 when referring to the inshot variable-load valve.

With an unladen vehicle, the inshot aspect reacts when the output from the foot valve is below the inshot design pressure of the variable-load valve, by allowing the supplied air to enter chambers (15), (38), (43) and (23). This pressure introduced into chamber (23) produces the inshot feature by requiring a higher pressure in chamber (28) to balance the downward force of piston (3). Thus, piston (3) will rise, closing inlet valve (16) on seat (17), but without opening the exhaust route (21). Therefore, the input and output pressures of this valve will be matched, until maximum inshot pressure is reached. Air pressure also flows from chamber (43) via passage (47) and balance pipe (48) to force piston (36) to rise and clamp ball pivot (31) against stem (20), in the same way as in a regular valve.

Beyond inshot pressure the main part of this valve will operate as a regular variable-load valve. However, the increased pressure entering chamber (43) causes piston (44) to compress inshot spring (45). Inlet/exhaust valve (40) is sealed against seat (42), closing the route from chamber (43) to chamber (38), ensuring that air pressure in chambers (43) and (23) cannot exceed the inshot design pressure.

When pressure rises in chambers (15) and (38), secondary piston (40) unseats piston (44) away from inlet/exhaust valve (39). Pressure in chambers (43) and (23) escapes to atmosphere via the slot in the end of inlet/exhaust valve (40), the hollow stem of piston (44) and exhaust flap (46).

Load-sensing valve MK1 (CD)

The load-sensing valve, incorporated in the air pressure braking system, is a device which automatically produces the most satisfactory brake force distribution to suit the load being carried by the vehicle. The valve is mounted on the chassis side member adjacent to the rear axle and connected to the axle by special linkage. A shock absorber and a knuckle joint are incorporated in the linkage to reduce movements caused by rough road surfaces. The load-sensing valve is connected by air lines to the brake chambers on the axle and to the air delivery ports in the footbrake valve. During a brake application, the load-sensing valve receives compressed air from the brake valve and delivers the air to the brake chambers on the axle at a pressure which is determined by the load being applied to the brake treadle by the driver.

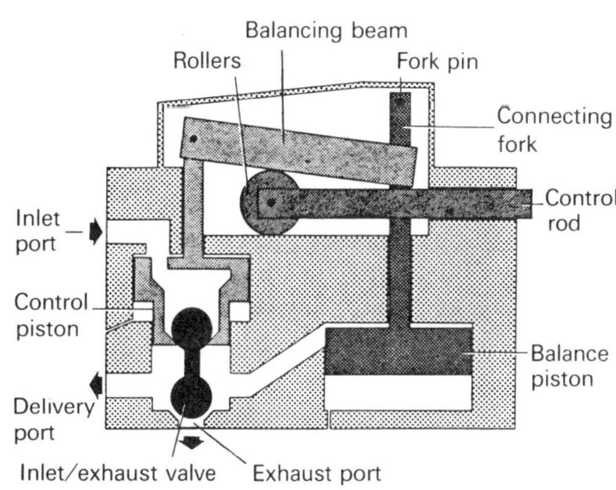

Figure 6.113 Load-sensing valve MK 1 (principles) (CD)

As shown in Figure 6.113, the load-sensing valve consists of a control valve assembly, a balance piston assembly contained in a housing, and a housing and cover containing a control rod assembly and a balance beam.

The control valve assembly comprises a body containing a piston which incorporates a spring-loaded inlet/exhaust valve assembly. A connecting rod assembly, secured to the piston by a retainer and two circlips, operates in a bore in the housing and is attached to one end of the balance beam.

The balance beam rests on two rollers carried on a pin held in a slot at one end of the control rod. The pin carries two large rollers which move along a steel plate, relative to the load on the vehicle at that time. The balance piston housing is connected to the control valve body by a transfer tube,which is fitted with two sealing rings to prevent air leakage. The balance piston is biased towards the top of the housing by a spring located in the housing. A connecting fork, secured to the top of the piston, passes on each side of the control rod and balance beam. The fork is fitted with a pin which contacts the balance beam during operation of the unit.

To prevent dirt and moisture from entering the unit, the exhaust port in the control valve body and the breather port in the balance piston housing cover are fitted with filter elements. The housing and bush assembly is fitted with a gaiter to protect the sliding surfaces of the control rod and bush.

Operation: The pressure of the air in the brake chambers controlled by the load-sensing valve depends upon the pressure of the air in the control air line from the foot valve and the position of the rollers in the load-sensing valve. The position of the rollers reflects the load being carried by the vehicle. When the vehicle is fully laden, the rollers are situated adjacent to the balance piston fork. The valve is so designed that when the rollers are in this position the balancing action of the valve does not occur, and the unit merely maintains the air in the brake chambers at the control air line pressure.

Application: When the driver depresses the foot valve treadle, compressed air enters the load-sensing valve and moves the control valve piston downwards. When this occurs, the exhaust valve closes the exhaust passage and the inlet valve is lifted off its seat in the control valve piston. Compressed air then passes through the unit to the brake chambers and through the transfer tube to the cavity above the balance piston. As compressed air is being delivered to the brake chambers, pressure is built up in the cavity above the balance piston. The pressure moves the balance piston and connecting fork downwards causing the fork pin to contact the balance beam.

Balancing: As shown in Figure 6.114, when the vehicle is unladen, or loaded to any point below maximum, the connecting fork pin depresses the end of the balance beam which raises the control valve piston until the inlet valve reseats and cuts off the air supply to the brake chambers and to the cavity above the balance piston. The unit is then in a balanced condition, with the inlet and exhaust valves seated, and with the forces acting on the balance piston balancing the forces acting on the control valve piston.

Should the driver depress the foot-treadle further and increase the air pressure above the control valve piston, the unit will increase the pressure in the brake chambers until a new point of balance is reached. If the driver reduces the effort applied to the foot-treadle, the air pressure above the control valve piston will be reduced and the balance beam, assisted by the air pressure in the cavity below the inlet valve, will lift the control valve piston. When this occurs, the inlet valve remains seated but the exhaust valve opens.

Air from the brake chamber and from the cavity above the balance piston will then pass through the exhaust filter element to atmosphere. As soon as the pressure above the balance piston has been reduced sufficiently, the control valve piston moves downwards and reseats the exhaust valve, preventing further loss of air from the brake chambers. With both valves seated, the unit is again in a balanced condition.

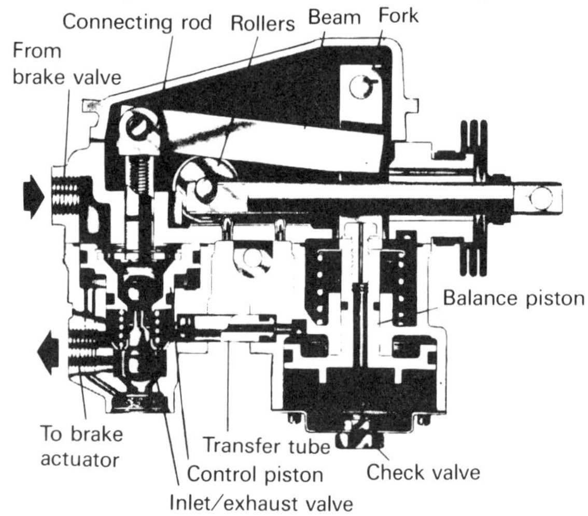

Figure 6.114 Load-sensing valve MK 1 (CD)

Releasing: When the foot-treadle returns to the brake fully-released position, the air pressure above the control valve piston is exhausted to atmosphere through the foot valve. When this occurs, the control valve piston is lifted by the balance beam and by air pressure in the cavity below the inlet valve. As soon as the control valve piston is lifted, the exhaust valve opens and all air pressure from the brake chambers and from the cavity above the balance piston, is exhausted to atmosphere and the brakes are fully released. When the brakes are fully released, the balance piston is raised to the top of its stroke by the return spring.

Load-sensing relay emergency valve (WO)

As shown in Figure 6.115, air enters port (1) via the red emergency (supply) line, and passes through one-way seal (p) and port (1-2) en route to the trailer reservoir. Should a drop in air pressure occur in the red emergency (supply) line, piston (q) will rise, closing exhaust port (R), and opening inlet port (k), to provide a through route from the trailer reservoir to the trailer brake chambers via ports (1-2) and (2).

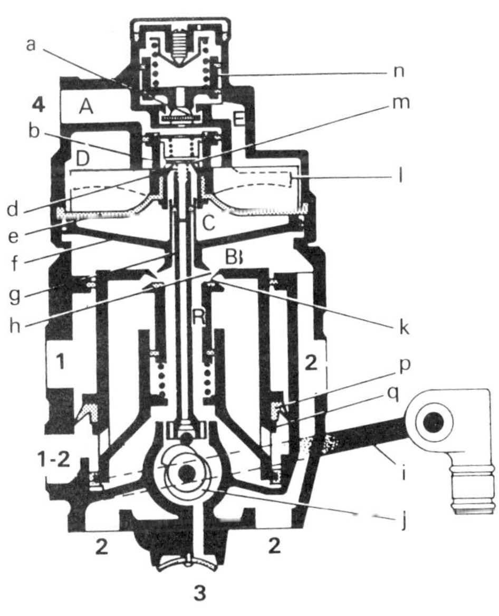

Figure 6.115 Load-sensing relay emergency valve (WO)

This valve is mounted on the trailer chassis and arm (i) is connected to the axle by linkage. With the trailer unladen the arm is at its lowest position and when fully laden the lever is in the highest position. The lever is connected to cam (j), which rotates and moves tappet (g) according to the load carried.

When the foot valve on the drawing vehicle is applied, air is supplied to the yellow service line that is connected to port (4) and chamber (A). This causes piston (b) to move down, closing exhaust port (d) and opening inlet port (m). Pressure also enters chamber (C) to act upon piston (f).

At the same time, air enters open valve (a) and passage (E) to reach chamber (D), and act on the top surface of diaphragm (e). This pressure is sufficient to prevent pulsing of pressure at low-control pressures (up to a maximum of 1 bar). When control pressure increases, piston (n) rises and closes valve (a).

As a result of an increase of pressure in chamber (C), piston (f) moves down, closing exhaust port (h) and opening inlet valve (k). Pressure from the trailer reservoir passes through port (1-2), port (k), chamber (B) and port (2), to reach the brake chambers.

Increasing air pressure in chamber (B) acts beneath piston (f), causing it to rise and close port (k).

The upper surface of diaphragm (e) is partially supported on its upper surface by the finned ring (l). When the force acting on the underside of the diaphragm (e) is equal to the force acting on piston (b) the piston moves up closing port (m), and a lapped condition is obtained. The position of valve tappet (g), which is controlled by lever (i), determines the lapped setting as the area of the diaphragm is modified.

A drop in service-line pressure acting at port (4) causes pistons (f) and (b) to move upwards, as result of the pressure in port (2) and chamber (C) respectively. This opens exhaust ports (d) and (h) allowing air to vent to atmosphere via exhaust (3).

Automatic load-sensing valve for air suspension (WO)

With a vehicle or trailer that is fitted with air suspension, the load cannot be 'sensed' by the relative distance between chassis and axle, as this remains almost constant with this type of suspension. To overcome this problem, the load-sensing valve used, reacts to the changes in air pressure that occur within the air suspension bellows when the load is changed.

This valve is mounted on the vehicle chassis. As shown in Figure 6.116, ports 41 and 42 are connected to the bellows of the air suspension. Pilot pressure from those bellows acts against pistons (m) and (k), and also moves guide sleeve (i) and the cam (h), against the resistance of spring (z). The profile presented by the cam depends upon the amount of the control pressure and the force of spring (z).

When the service brake is applied, compressed air

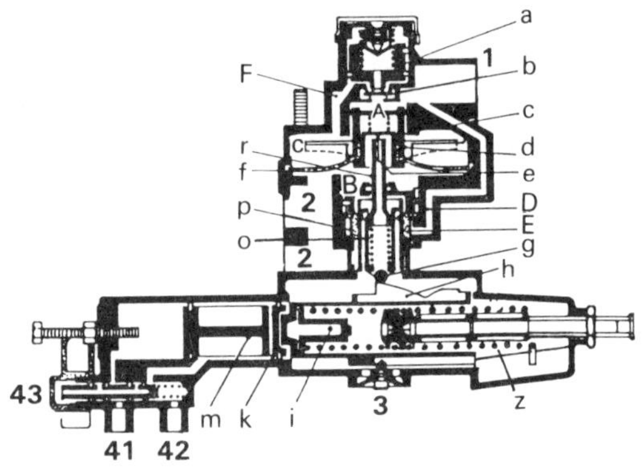

Figure 6.116 Automatic load-sensing valve (WO)

from the relay emergency valve, enters port (1) and chamber (A), to move piston (d) down, closing outlet (e) and opening inlet (c). Air can then enter chamber (B) beneath diaphragm (f), and pass to the brake chambers, via ports (2).

Simultaneously, air enters chamber (C) through valve (b) and passage (F) to act down on diaphragm (f). This provides an inshot of air when the footbrake treadle is only lightly applied. When the pilot pressure is increased, piston (a) is moved up and valve (b) closes.

Note: The term 'inshot' means supplying a shot of compressed air. This is achieved by matching the pressure output of the load-sensing valve to the pressure output of the foot valve, up to a maximum preset figure of around 10 lbf/in^2 (70 kN/m^2, 0.7 bar), depending upon the particular model and subject to the manufacturers' recommendations. The inshot of air, when only a light brake application is made, ensures that sufficient air is supplied to overcome the stiffness of the foundation brakes, and to avoid glazing the lining material. Above that figure, the pressure delivered to the brake chambers, by the load-sensing valve is again relative to the load on the controlled axle of that vehicle.

As piston (d) moves down, diaphragm (f) gradually accepts the bell shape of the underside of piston (d). The effective area below this diaphragm progressively increases until it is equal to that on the upper side of piston (d). This results in piston (d) rising and closing inlet (c). Thus, with both inlet (c) and the hollow exhaust valve stem (r) closed, the brakes remain applied and the load-sensing valve is in a lapped, or balanced, condition. However, when the vehicle is fully-laden, and the footbrake treadle is applied, the pressure supplied to the load-sensing valve, is passed on to the brake chambers, without reduction.

When the footbrake treadle is released, air pressure in chamber (B) causes piston (d) to move up, opening exhaust (e) and allowing air to escape to atmosphere via hollow valve stem (r) and exhaust port (3).

Whenever the footbrake treadle is applied, compressed air from port (1) of the load-sensing valve, passes through airway (D) to enter chamber (E) to act upon flexible clamping ring (p). This causes frictional resistance between valve stem (r) and the housing. This prevents variations in the valve ratio during braking, due to changes in pressures within suspension bellows. To allow for roller (g) moving upwards during dynamic changes, spring (o) compresses, rather than causing valve stem (r) to move.

Levelling valve for air suspension (WO)

This valve charges and exhausts the bellows of a pneumatic suspension system, in order to maintain a constant chassis height, above the road surface. The piping layout is shown in Figure 6.117.

As shown in Figure 6.118, when the load on a chassis is increased, the vehicle chassis descends, i.e. it gets closer to the axle. The levelling valve is connected by linkage from lever (8) to the axle. As the chassis goes down, lever (8) is pushed up, moving eccentric peg (6), and pushing follower (7) and stem (4) upwards, opening inlet valve (2).

Reservoir pressure at port (V) opens check valve (1), allowing compressed air to pass out of delivery port (L) to the air bellows.

To minimise air consumption, the top of stem (4) has a grooved pattern that can give two different cross-sectional areas, so that the through route is governed by the amount that lever (8) is moved.

This valve will reach a lapped, or balanced, condition, when the suspension bellows have received the correct pressure and the chassis is in such a position that lever (8) is horizontal, causing inlet valve (2) to close.

When the load (or part of that load) is removed from the vehicle, the reverse of the above process occurs. As the pressure in the bellows is now excessive for the lesser load, the chassis moves up and lever (8) moves down. Stem (4) moves down, allowing inlet valve (2) to reseat, and for the excess pressure to exhaust to atmosphere via central exhaust bore (3) and exhaust port (E) as pressure normalises. The chassis and lever (8) both regain their horizontal position. The valve again resumes a lapped, or balanced, condition, with stem (4) sealing against the closed inlet valve (2).

To prevent this valve responding to the wheel

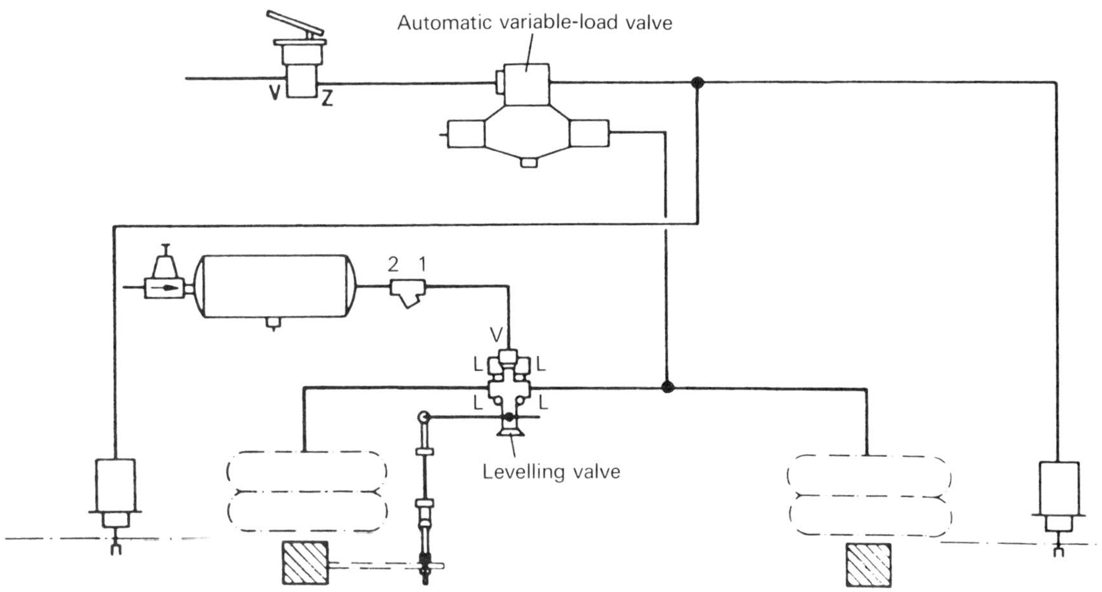

Figure 6.117 Air suspension piping layout for automatic load-sensing regulation

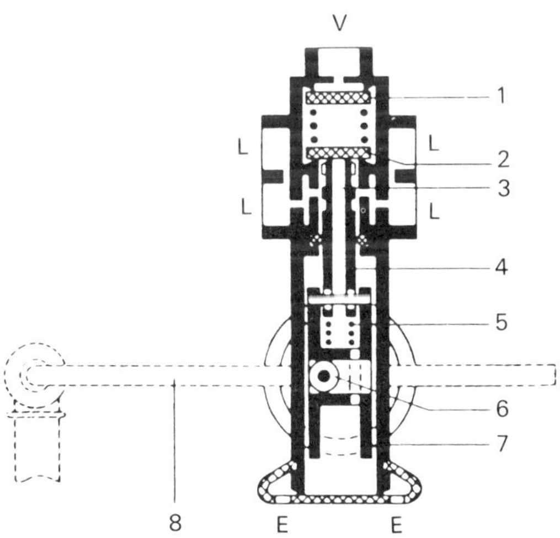

Figure 6.118 Levelling valve (WO)

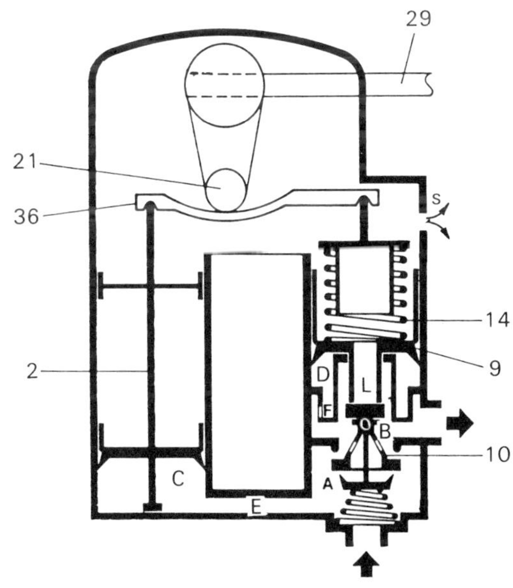

Figure 6.119 Load-sensing valve (BH)

bounce that is encountered when crossing rough surfaces, a limited amount of free-play is built into this unit. Therefore, limited deflection of lever (8) will move follower (7), but spring (5) will absorb the shock and prevent movement of stem (4).

Load-sensing valve (BH)

The function of this unit is to automatically vary the amount of braking on a particular axle (or axles) according to the load being carried on that axle (or axles).

As shown in Figure 6.119, this valve is attached to the chassis and connected to the axle by lever and push-rod linkage. The distance between the chassis and axle decreases as the load carried is increased and vice versa. It is the distance between these two items that alters the angle of lever (29). At maximum load this lever is inclined upwards whereas in the unladen position it is inclined downwards. Chamber (A) is connected to the foot valve. When the foot valve is applied, compressed air enters chamber (A) and passes through chamber (B) leaving the valve via the delivery port, en route to the brake chambers.

Simultaneously, air pressure flows through passage (E) to chamber (C), where it acts upwards upon piston (2) which is connected, at the top, to balance beam (36). The position of the balance beam, in relation to the horizontal, is determined by the position of roller (21), which acts as a fulcrum for the balance beam (36), and which itself is governed by the position of lever (29) that relates to the load being carried on the vehicle.

Unladen condition: If the vehicle is only carrying a light load, the chassis is at its highest possible position in relation to the axle. The lever (29) takes up an acute downwards angle, and the fulcrum roller (21) is at the extreme left side of the balance beam. Air pressure being delivered to the brake chambers is only about one-third of the input from the foot valve, due to the light load upon spring (14).

Fully-laden condition: When the vehicle is fully-laden, the chassis is at a comparatively low position in relation to the axle, and lever (29) is at an acute upward angle, whilst the fulcrum roller (21) is at the centre of the balance beam. The input and output pressures from this valve are now identical.

Balanced condition: When the air pressure acting upwards beneath the control piston (9), exceeds the force of graduating spring (14) acting downwards, the whole piston assembly will lift closing inlet valve (10). The valve will then be in the lapped, or balanced, condition. The brakes will remain held on, as long as the foot valve remains applied. When the foot valve is released compressed air will cease to supply chambers (A), (E) and (C) and the compressed air will escape from the brake chambers, to atmosphere, via hollow stem (L) and exhaust port (S).

Load-sensing valve (PD)

This unit is mounted on the chassis and connected, via a moveable arm and rubber linkage, to the axle concerned.

As shown in Figure 6.120, in the normal driving condition, i.e. brakes not in use, graduating piston (8) is held at the right-hand end of its bore by compression spring (9). At this stage, outlet port (10) is closed and inlet port (11) is open.

When the driver applies his foot valve treadle, compressed air from that valve enters the load-sensing valve, through inlet port (1), passing through, and out of delivery port (2), en route to the brake chambers.

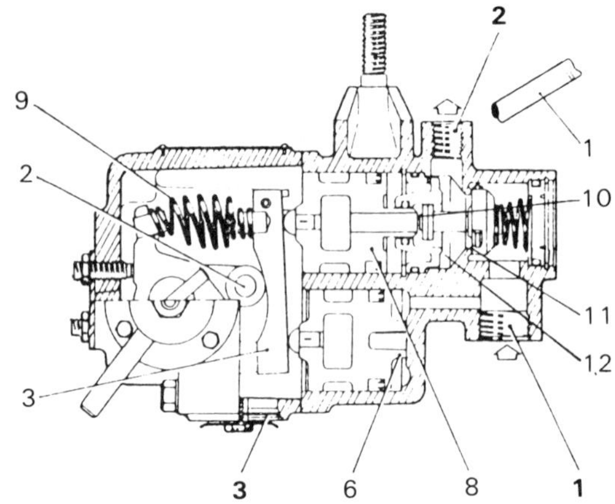

Figure 6.120 Load-sensing valve (PD)

A low air pressure does not activate this valve. However, when the pressure rises above approximately 7.25 lbf/in^2 (50 kN/m^2, 0.5 bar), sufficient to operate brake chambers, the compressed air moves both control piston (6) and graduating piston (8) to the left of their respective bore.

The pressure then delivered to the brake chambers depends upon the position of adjusting lever (1), which is connected via a rubber link, to the vehicle's axle. The position of this lever is affected by the vehicle's load, and it acts at the end adjacent to the load-sensing valve, via roller (2), to cause a cam action upon balance lever (3).

The base of the balance lever (3) is moved to the left by control piston (6), by pivoting about point (2), this moves the top of the balance beam to the right, against graduating piston (8), moving it to the right, and opening inlet (11).

When the output air pressure, acting upon graduating piston (8), via pressure compensating bore (12), exceeds the force of compression spring (9), the piston moves to the left closing the inlet valve. The unit is then in the lapped, or balanced, condition, i.e. inlet and exhaust valves are closed and brakes applied. This situation will not change until there is an increase or decrease in the force applied by the driver's foot upon the footbrake treadle.

When the footbrake treadle is released, pressure above the control piston (6) is released, inlet valve (11) is closed and outlet (10) is opened, allowing the compressed air from the brake chambers to escape to atmosphere, via exhaust port (3).

Empty-load valve (WO)

The role of this valve is to regulate the output pressure, relative to the actuating pressure. In addition it

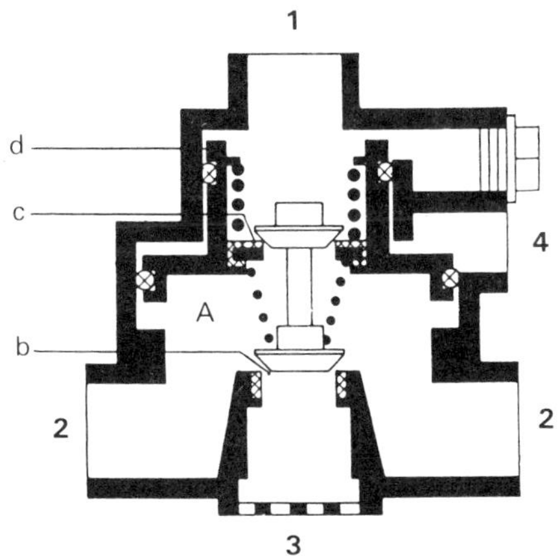

Figure 6.121 Empty-load valve (WO)

provides for the rapid exhaust of downstream appliances.

As shown in Figure 6.121, compressed air enters port (4) and acts on the horizontal surface of piston (d) moving it downwards, closing exhaust valve (b) and opening inlet valve (c). The supply pressure at port (1) then passes through the valve, leaving via delivery port (2), to the downstream appliances. Simultaneously, pressure builds up in chamber (A) and acts on the underside of piston (d). Inlet (c) is closed when input pressure at port (4) and output pressure at port (2) match the area ratio of piston (d).

When pressure drops at port (4), the increased pressure in chamber (A) moves up piston (d). Exhaust valve (b) opens and the downstream appliances are exhausted via exhaust port (3).

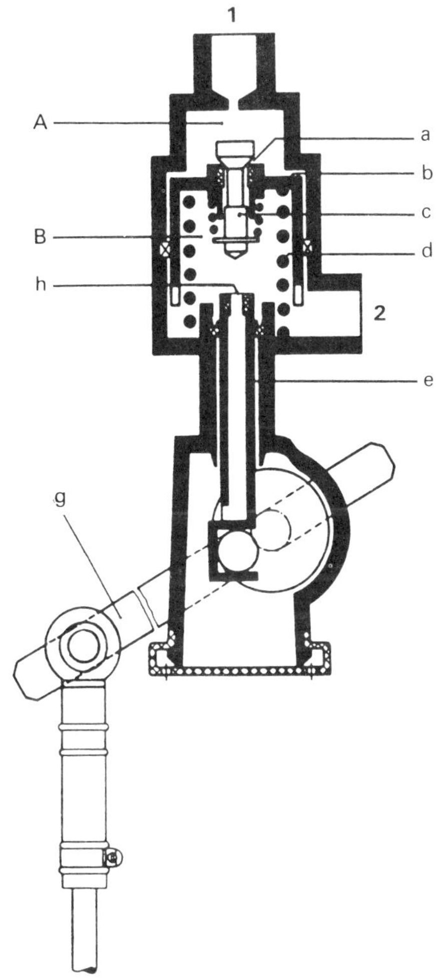

Figure 6.122 Load-sensing valve (WO)

Load-sensing valve (WO)

The function of this unit is to control the steering stability of a trailing axle, relative to spring deflection, and therefore vehicle load.

As shown in Figure 6.122, this valve is mounted on the chassis and connected to the axle via lever (g) and linkage. Lever (g) is at its lowest position when the vehicle is unladen, and the lever reaches its highest position when the vehicle is fully-laden.

Compressed air from the reservoir enters port (1), and chamber (A), pushing down piston (b) against return spring (d). Valve assembly (c) closes exhaust valve (h), which seals stem (e) and opens inlet (a). Air then flows to downstream appliances via chamber (B) and port (2).

Simultaneously pressure builds up in chamber (B) and acts on the underside of piston (b), causing it to lift. This closes inlet (a) and achieves a lapped, or balanced, condition.

Manual light-laden valve (WO)

This type of valve, fitted to a trailer, enables a driver to alter the amount of braking force available, by merely turning a lever on that valve, prior to driving on a particular journey.

As shown in Figure 6.123, when a brake application is made, compressed air enters, via the relay emergency valve, through port (S), into chamber (a), then via open inlet valve (4), out via port (Z) to the trailer brake chambers. At the same time, the pressure also acts upon the top face of piston (5), which is kept at TDC by return spring (6).

Prior to a journey, lever (7) is manually moved to one of three driving positions (full-load, half-load,

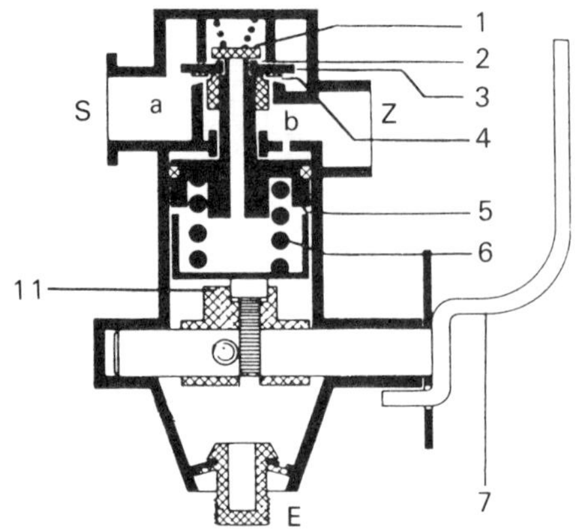

Figure 6.123 Manual light-laden valve (WO)

unladen), which moves eccentric cam (11) to preload spring (6).

When the air pressure acting upon the top of piston (5), overcomes the load in spring (6), it moves down, together with valve (3) and spring-loaded valve (1). Inlet valves (2) and (4) close, preventing any increase in pressure entering this unit and the brake chambers.

If there is a loss of pressure at the brake chambers and/or connecting line, spring (6) will move piston (5) upwards against spring-loaded piston (1), opening the inlet valve (2) to restore the lost pressure.

When the brakes of the towing vehicle are released, air pressure at port (S) and chamber (a) drop to zero. Consequently, the higher pressure in chamber (b) lifts valve (3) off its seat. Similarly, spring-loaded valve (1) is unseated and inlet valve (4) opens, allowing the brake chambers to exhaust to atmosphere via port (S) and the relay emergency valve. As air pressure is released from the above piston (5), it is returned to TDC by spring (6).

When the trailer is uncoupled, the trailer brakes can be released by turning lever (7) to the release position. Cam (7) removes the load from spring (6), allowing piston (5) to drop to BDC, opening exhaust valve (1). This enables compressed air in the brake chambers to escape to atmosphere via the hollow stem of piston (5) and exhaust port (E).

Hope anti-jack-knife device

Articulated vehicles are extremely popular because of their ability to drop a laden trailer and pick-up an unladen trailer with a minimum of time loss. They are also highly manoeuvrable, but their stability is severely lacking when compared with a rigid vehicle. An articulated trailer is far longer than its drawing unit; the king pin (pivot) is at the extreme front of the trailer, whereas the axles are at the rear. In an ideal, straight-ahead direction, these factors aid stability, but when any lateral movement occurs, these features contribute to instability. On a wet/greasy surface, an unladen articulated vehicle needs careful handling. Fierce braking can induce a jack-knife or alternatively trailer swing (see Figure 6.124).

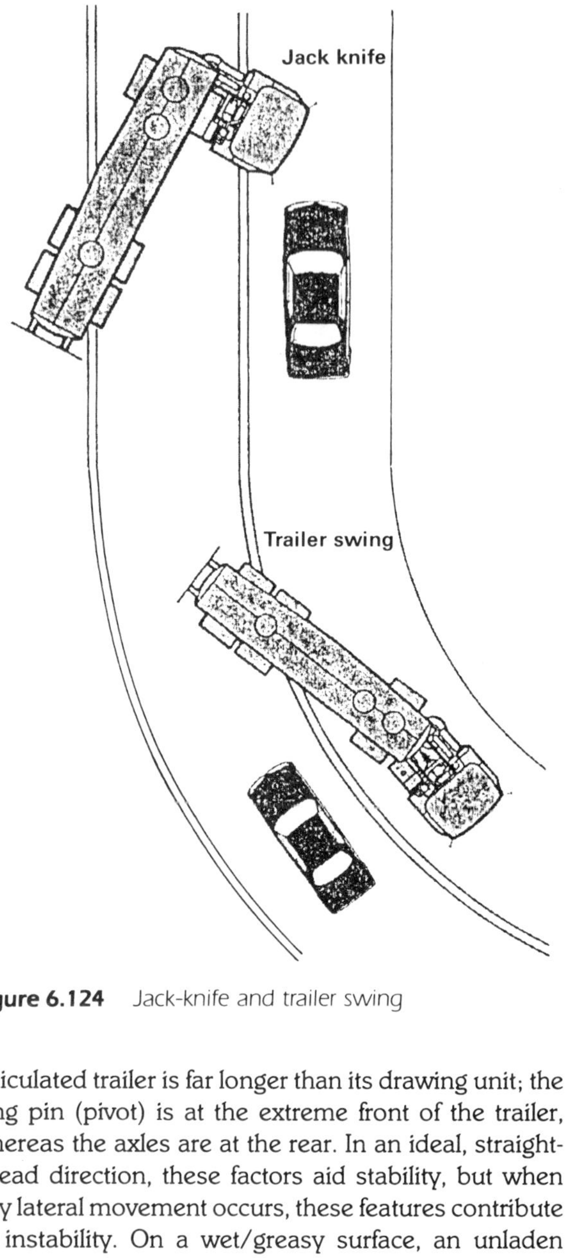

Figure 6.124 Jack-knife and trailer swing

The Hope anti-Jack-knife device (see Figure 6.125) assists in preventing this excessive instability by providing friction damping at the pivot point. The level of damping is automatically controlled by the action of the foot or handbrake valve. This device does not comply with all the EEC requirements for anti-lock braking systems. However, certain large haulage fleets have fitted the Hope device, in addition to electronic anti-lock, to restrict lateral movement when braking.

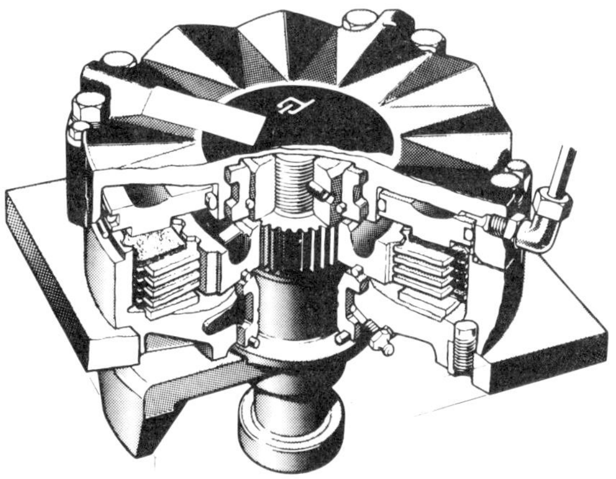

Figure 6.125 Hope anti-jack-knife device (HE)

This unit consists of a multi-plate friction clutch, which is mounted on the end of the king pin. The king pin is free to turn within a sealed body assembly which is fixed to the trailer. An extended arm on the king pin forging locates in the throat of the fifth wheel coupling and acts as the drive to the four steel discs, which are located by splines on a gear attached to the top of the king pin.

Five friction discs, positioned between the steel discs, and located by splines within the main body, complete the friction pack. An air-operated piston, located in the cover of this unit, acts upon the discs in response to signal air pressure from the braking system.

The operation of the control valve is as follows: at 5 lbf/in^2 (35 kN/m^2, 0.35 bar) brake pressure, the valve delivers 40 lbf/in^2 (275 kN/m^2, 2.8 bar) to the device. This pressure remains constant until brake pressure reaches 40 lbf/in^2 (275 kN/m^2, 2.8 bar), when it will increase in unison with brake pressure to a maximum of 65 lbf/in^2 (450 kN/m^2, 4.5 bar) at the device, but allowing brake pressure to exceed this amount as required.

Anti-lock

Electronic anti-lock devices have been incorporated into the braking systems of military aircraft and commercially-operated, passenger-carrying aircraft for many years. Such aircraft require fast landing speeds, a limited distance in which to stop, a variety of surface textures on runways, plus a complete range of climates and weather conditions. It sounds like the perfect test conditions necessary to prove the serviceability for any type of equipment.

Since the 1960s brake manufacturers and governments have been trying to convince hauliers and public service vehicle operators that electronically-controlled, anti-lock braking systems are an important aspect of road safety.

There were early problems associated with external interference by citizen band (CB) and emergency radios, but these have been resolved. Indeed, a current EEC Directive states that the operation of anti-lock systems must not be adversely affected by magnetic or electrical fields.

The EEC have also decided that all new vehicles of certain types will be required to have anti-lock braking equipment, by a date that has yet to be finalised. The following vehicles will require a Category 1 anti-lock system, i.e. the braking on all road wheels will be individually sensed:

1 Category M3 Motor Vehicles, i.e. inter-urban motor coaches and long-distance touring coaches, but at present *only* those with a gross weight in excess of 12 tonne.
2 Category N3 Motor Vehicles, i.e. heavy goods vehicles, but at present *only* those having a gross weight exceeding 16 tonne, authorised to tow Category 04 trailers, i.e. those having a gross weight exceeding 10 tonne.

The following will require an Annex X anti-lock system, i.e. the braking will be sensed on *each* axle:

3 Category 04 trailers, i.e. full, semi and centre trailers with a gross weight exceeding 10 tonne.

Note: There are vehicles that come under the classification of Category M3 and N3, which are below the weights shown above; these are *not* required to fit anti-lock systems.

For the immediate future, operators of vehicles and trailers *outside* the specified groups will not have a legal obligation to have anti-lock braking systems fitted, although this exemption will probably only remain for a limited period, depending upon the success or otherwise of this initial obligation.

Early anti-lock systems for road vehicles were known as anti-jack-knife devices, as they were fitted to the drawing unit of an articulated vehicle to sense the lock-up of the road wheels on the drive axle. Trailer systems were also introduced, where only one wheel on each axle was sensed. However, the latest systems are now sensing each wheel on rigid vehicles, drawing units and trailers.

The use of electronics to sense when the braked

road wheels of a vehicle and/or trailer are about to lock, and then to reduce the braking effort at those wheels, thereby averting a skid, must be a major contribution to road safety. This system is merely automating the well-known principle of cadence braking which is practised by rally drivers, and indeed taught to bus drivers, to minimise the risk of skidding on slippery surfaces. The principle of this system is that the driver quickly dabs the footbrake pedal repeatedly 'on' and 'off', so that the road wheels are released before they lock. Maximum braking efficiency is achieved just before the wheels lock, but once a road wheel locks, efficiency drops dramatically.

Wheel lock is an undesirable feature associated with orthodox braking. Its effects include excessive tyre wear, instability, loss of steering control, and in the case of articulated vehicles, jack-knifing or trailer swing. The results of such malfunctions can range from extra running costs to a major road accident. In 1988, in the UK the following accidents, due to skidding, were reported: heavy goods vehicles 16 376; buses or coaches 12 086.

Skidding can be the result of harsh braking, causing the road wheels to lock-up, but it can also be due to lesser braking on a *split-mew* surface. (*Mew* is the term used for the coefficient of friction, and a *split-mew* surface refers to having parts of the road with a good abrasive surface and other parts giving less adhesion to the tyre.) A typical example is a road with part of its width covered in diesel oil or ice, but with the rest of the surface presenting a good abrasive braking surface. On a split-mew surface, careful braking is required to prevent skidding and to maintain vehicle stability. Anti-lock braking systems achieve these objectives automatically.

Electronic braking control systems are manufactured by a number of companies, including:

Bendix ABS System
Bosch ABS System
Grau Skidcheck
Wabco ABS System.

A Bosch ABS System is shown, as one example, in Figure 6.126.

In all cases the anti-lock systems are made up from several essential component parts, and although they are often given different names or produced as combined units, and may differ in mode of operation, they control the following functions:

1 to *sense* road wheel speed;
2 to *monitor* that speed;

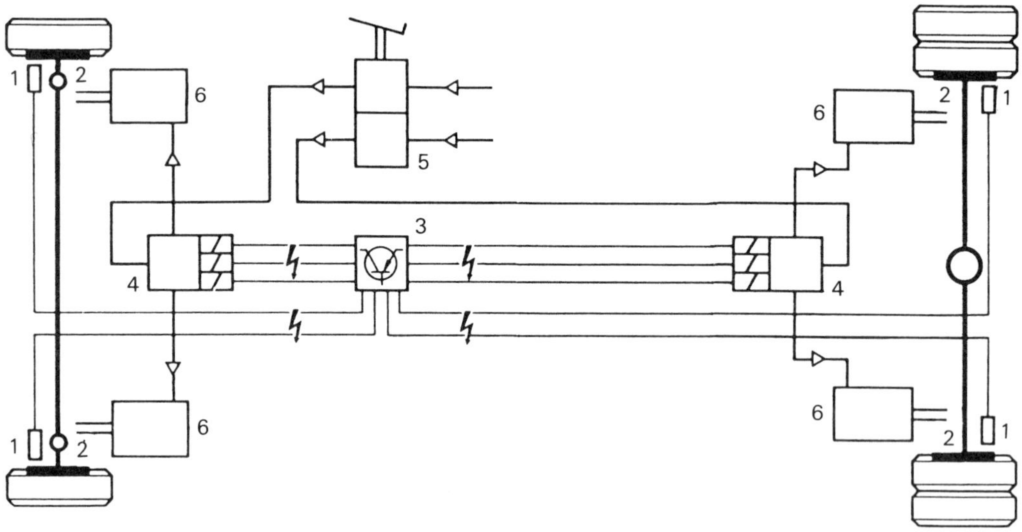

Key
1 = Wheel spread sensor
2 = Exciter ring
3 = Electronic controller
4 = Pressure-control valve
5 = Foot valve
6 = Brake chamber

Figure 6.126 Layout of 4 × 2 commercial vehicle with ABS (BH)

3 to *command* correction;
4 to *modulate*, i.e. to release and reapply the brakes as necessary;
5 to give *warning* to the driver, first, when starting-up, that the correction system is operational and second, when driving, if the system fails.

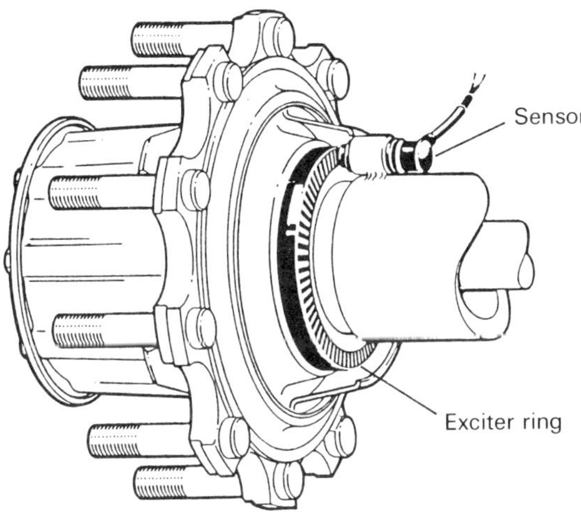

Figure 6.127 Exciter ring and sensor

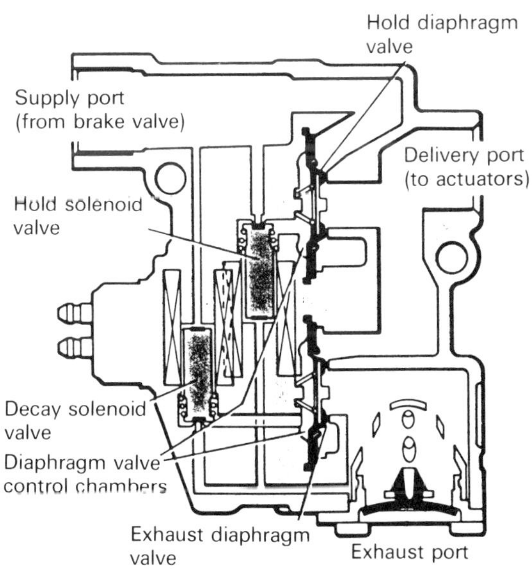

Figure 6.128 Modulator valve (BW)

Figure 6.127 shows an exciter ring and sensor, and Figure 6.128 shows a modulator valve. These are two examples of component parts used to control the above functions.

Additionally, three of the above manufacturers are offering systems that incorporate a traction control system that enables a vehicle to be driven, even if a road wheel starts to spin on a split-mew surface.

Figure 6.129 shows the layout of a 4 × 2 truck or bus system with ABS and ASR.

To *sense* the road wheel speed, all of these systems fit a slotted exciter ring to the hub of each road wheel, irrespective of whether the vehicle is the drawing unit of an articulated vehicle, a rigid vehicle or a trailer.

To *monitor* the road speed, there is a sensor fitted close to each exciter ring. As the road wheels turn, the exciter rings generate pulses in the stationary sensors, which are proportional to the wheel speed.

Changes in these pulses are noted by the microprocessor, in the electronic control unit, which calculates the deceleration and *commands* the necessary correction.

The modulator valve (solenoid valve) receives the commands from the electronic control unit, which causes it to *modulate* the brake force, i.e. release and reapply the brakes as necessary.

Warning is given to the driver, by a lamp on the dashboard, that the system has failed. The same lamp also provides a check that the system is functional, when the vehicle is started up.

Modulator valve (BW)

This modulator valve (see Figure 6.128), which contains two solenoid valves, responds quickly to signals from the electronic control unit. One solenoid controls the air flow through passages to the delivery port, en route to brake chambers, and the other solenoid controls the air flow to the exhaust port of the modulator valve.

When normal braking takes place, the solenoids are not energised and air flows from the supply port through to the brake chambers, via the delivery port. Some of this air also flows through the open decay solenoid valve, to hold the exhaust diaphragm valve in the closed position.

When an anti-lock situation occurs, both solenoids are activated. The hold solenoid allows air to pass and so close the hold diaphragm valve, thereby closing off the route to the delivery port and the brake chambers. The decay solenoid opens and releases pressure from the exhaust diaphragm control chamber, which opens the exhaust valve, providing a direct route to enable the brake chambers to exhaust to atmosphere.

By simply activating the hold solenoid valve, air flow from supply to delivery is prevented, but the exhaust from the modulator valve is also closed, thus a constant

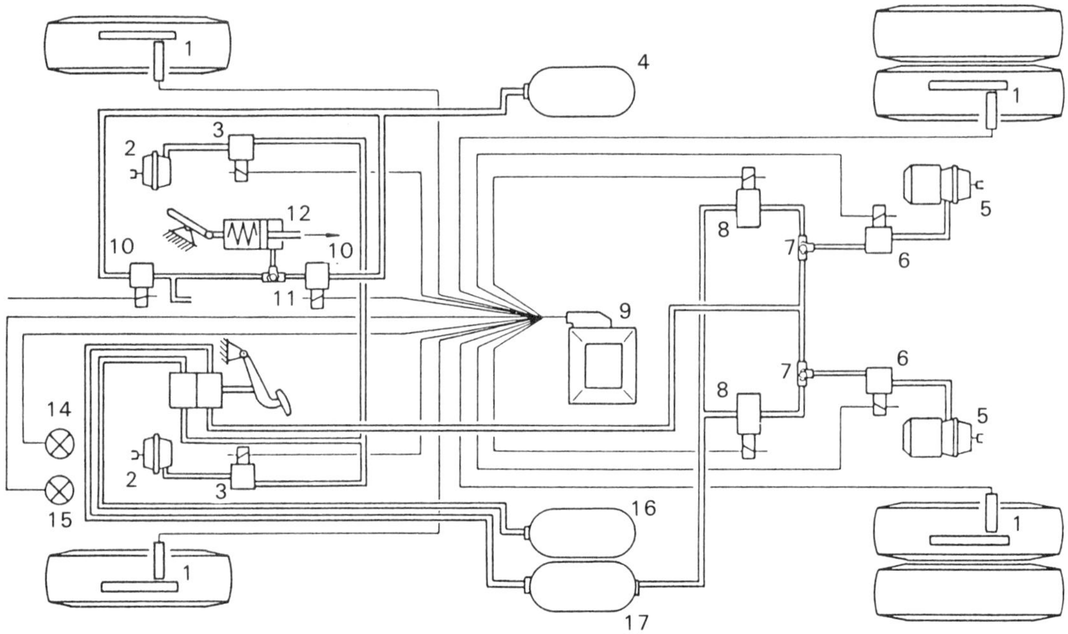

Key

1 = Wheel sensors (front wheels)	9 = Electronic control unit
2 = Front brake chambers	10 = Solenoid valve to control engine
3 = Solenoid control valve	11 = Double-check valve
4 = Auxiliary reservoir	12 = Drive slip control (ASR) cylinder
5 = Rear brake chambers	14 = ASR indicator lamp
6 = Solenoid control valve	15 = ABS warning lamp
7 = Double-check valve	16 = Reservoir
8 = Solenoid valve	17 = Reservoir

Figure 6.129 Layout of 4 × 2 truck or bus system with ABS and ASR (WO)

air pressure is retained at the delivery port and at brake actuators.

The electronic control unit is, as its name suggests, a totally electronic item which cannot be adjusted or repaired in a motor vehicle workshop.

Although individual wheel control is ideal for driven wheels, it has disadvantages on steered wheels when a vehicle is travelling on a split-mew surface. The vehicle would be difficult to handle and tend to *yaw* (slew to the side). Consequently, these systems incorporate a feature to control the differential pressure across the steered axle. When a front wheel starts to skid, the air pressure to both front brake chambers is held and then the pressure to the sliding wheel starts to decay until the wheel grips. The wheel on the opposite side has a short pressure decay initially, followed by a brakes-applied cycle. These cycles are repeated until normal braking is achieved.

As with other components described in this publication, it is not suggested that any one particular system or component is either superior or inferior to similar devices produced by other companies. The aim is merely to present the alternatives to the reader.

Drive slip control or traction control

The terms 'drive slip control' and 'traction control' are alternatives used by different manufacturers to distinguish their product from the product of another company. In both cases, the system is designed to enable a vehicle to be driven on slippery road surfaces without encountering the normal wheel spin problems.

These systems are incorporated into an anti-lock system and use the same foundation brakes to restrain a spinning wheel on a drive axle. When the orthodox differential gear is allowing a wheel to spin on a slippery road surface, the electronic control unit will cause the brake to be applied to the spinning wheel. Engine torque will automatically be increased to the nonslipping wheel by the action of the differential gear of that axle.

In a situation where both driving wheels are spinning, the engine torque will be automatically reduced until the spin ceases when a balance is achieved between a gradual release of the brakes and a gradual acceleration of the drive wheels of the vehicle.

A number of components are incorporated into an anti-lock system to achieve the addition of an anti-slip control system.

Differential brake valves

These valves are connected to the solenoid control valves (modulator valves) of the driven wheels and supply compressed air to the brake chamber of the spinning drive wheel as necessary.

Engine control valve

This valve causes the drive slip control cylinder to regulate the supply of fuel to the engine.

Drive slip control cylinder

This air cylinder, which is activated by compressed air from the engine control valve, reacts by moving the fuel injection pump control rack as necessary, according to the grip/slip of the road wheel.

Indicator lamp

As with ABS, all models of ASR have a warning lamp on the dashboard of the vehicle to indicate to the driver that the system is correct, when starting up the vehicle. These lamps also signal to indicate that a problem has occurred whilst the vehicle is in service.

Turn cock with exhaust (WO)

This unit caters for the charging and exhausting of hoses to a trailer, when that trailer is being coupled or uncoupled.

As shown in Figure 6.130, as lever (1) is turned this causes the eccentric cam (3) to open inlet valve (2). Thus, when lever (1) is parallel with the pipework, air flows through the turncock to the trailer. However, when lever (1) is turned through 90° (as with a gas tap), valve (2) is closed by spring (4) and the line between this unit and the trailer is exhausted.

C and CA-couplings (FJ)

The C and CA-couplings are in common use within the UK providing a simple, quick-release, leak-proof connection between the towing vehicle brake hoses

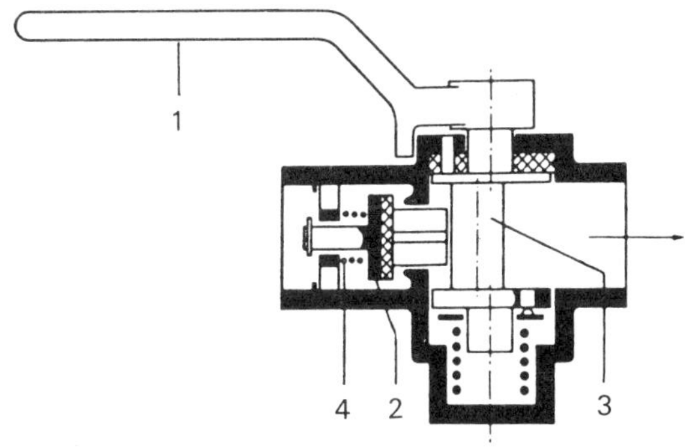

Figure 6.130 Turn cock with exhaust (WO)

and the trailer brake pipe line. They incorporate a self-sealing device that prevents air loss, when the towing vehicle is operating as a solo unit. An example of a C-coupling is shown in Figure 6.131.

To aid the driver and to prevent the cross-coupling of systems, it is common practice for C and CA-couplings to be fitted in the following patterns:

Tractor	*Trailer*
Service, control hose with male C-coupling	Female C-coupling
Secondary, auxiliary hose with female CA-coupling	Male CA-coupling
Emergency, supply hose with female C-coupling	Male C-coupling

To connect male and female couplings, the outer sleeve of the female section is rotated to locate an indent and thus allow it to be moved along the coupling body. The male section is then entered and the sleeve rotated to provide a safety key. To release the coupling the procedure is reversed.

A blanking plug or socket should be fitted at the rear of the drawing vehicle, to locate the open end of hoses, when they are not connected to a trailer. Obviously, the opposite type of coupling is required, but in such cases the internal probe of the holding plug or socket must be removed to prevent air leaks.

Some users remove probes from couplings that are fitted between towing vehicle and trailer. This necessitates a turncock being fitted in each line at the rear of the towing vehicle. This gives rise to the risk of leakage if the driver forgets to turn off the turncocks, and the lack of brakes on one or more of the trailer systems if the turncock is not activated.

Note 1: EEC Directives require that 'recent' type-approved vehicles *must* be fitted with self-sealing couplings, i.e. shut-off devices which are not automatically activated will not be permitted.

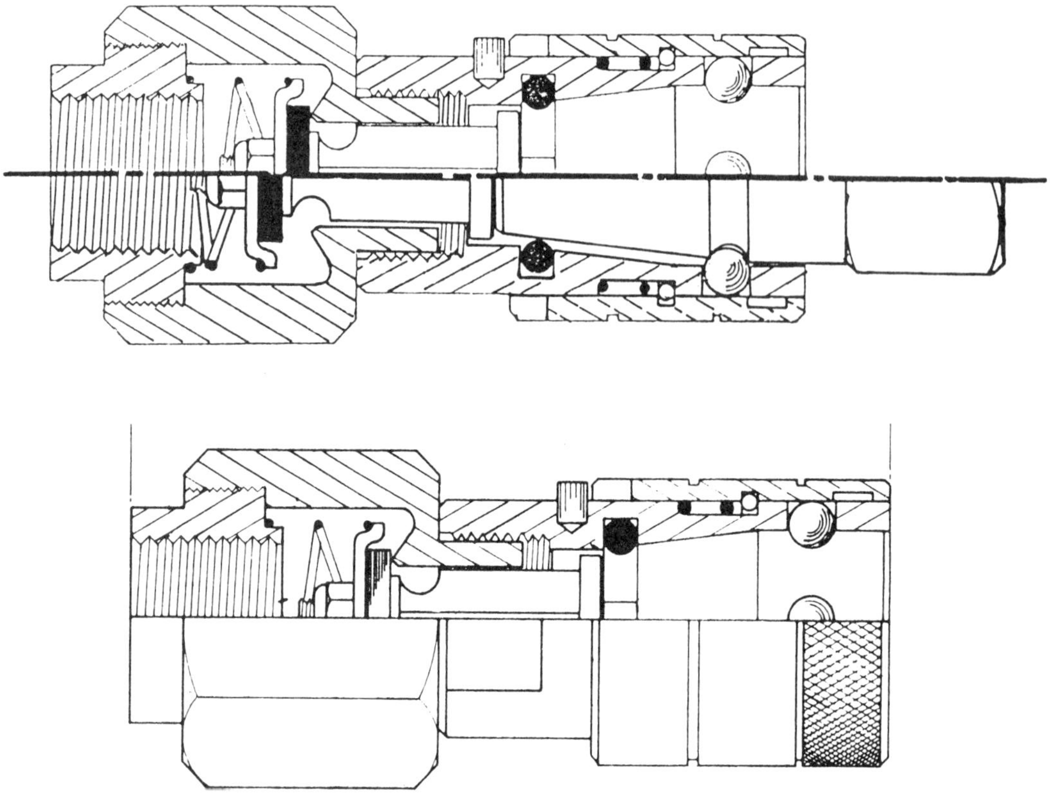

Figure 6.131 C-couplings (FJ)

Note 2: The Department of Transport have mounted an advertising safety campaign asking operators *not* to remove non-return valves from couplings.

Coupling heads – palm couplings (WO)

The role of the coupling head is to connect the air-brake system of a truck or tractor to the air-brake system of the trailer.

As shown in Figure 6.132, hose couplings of type A1, B1 and C1 serve the emergency (supply) line and all have a red cover. Types A2 and B2 serve the service (control) line and both have a yellow cover.

Note: All of these hose couplings have a device to prevent a wrong connection being made. Only types B and C are equipped with a built-in isolating valve which seals the air passage whenever the coupling head is left disconnected.

Connecting: To connect the trailer coupling head to the coupling on the tractor hose requires a turning action, after joining the two halves of the coupling. The full, secure and locked engagement of the coupling is when no further turning is possible. An anti-mismatch device is built into these components to prevent the cross-coupling of hoses.

Coupling of C1 to A1, B1 to A1 and B2 to A2: During coupling, the seal ring of the type A coupling opens type B or type C couplings, and also provides an air-tight seal. Upon disconnection, the valve will automatically seal.

Coupling A2 to A2: Identical couplings, without valves, achieve an air-tight seal by the compression of both seal rings one against the other.

Slack adjuster (BW)

Slack adjusters serve two purposes:

1 They act as a lever during normal braking.
2 They provide a quick and easy method of adjusting brakes and so prevent over-extension of the diaphragm.

Figure 6.133 shows an example of a slack adjuster while Figure 6.134 shows its component parts.

Slack adjusters use the worm and gear principle. Brake adjustment is made after depressing the locking sleeve and then turning the adjusting screw of the adjuster with a spanner so that it rotates the worm shaft. The worm shaft meshes with the worm gear which is connected to the brake cam, so by turning the worm gear the brake cam spreads the shoes and compensates for lining wear.

The slack adjuster should, after being properly adjusted, set the brake arm at just over 90° to the brake chamber push-rod *when the brake is applied*. Check that adjusting screws are locked after adjustment.

Automatic slack adjusters (HX)

There are several different types of automatic slack adjusters which have been in use for many years on

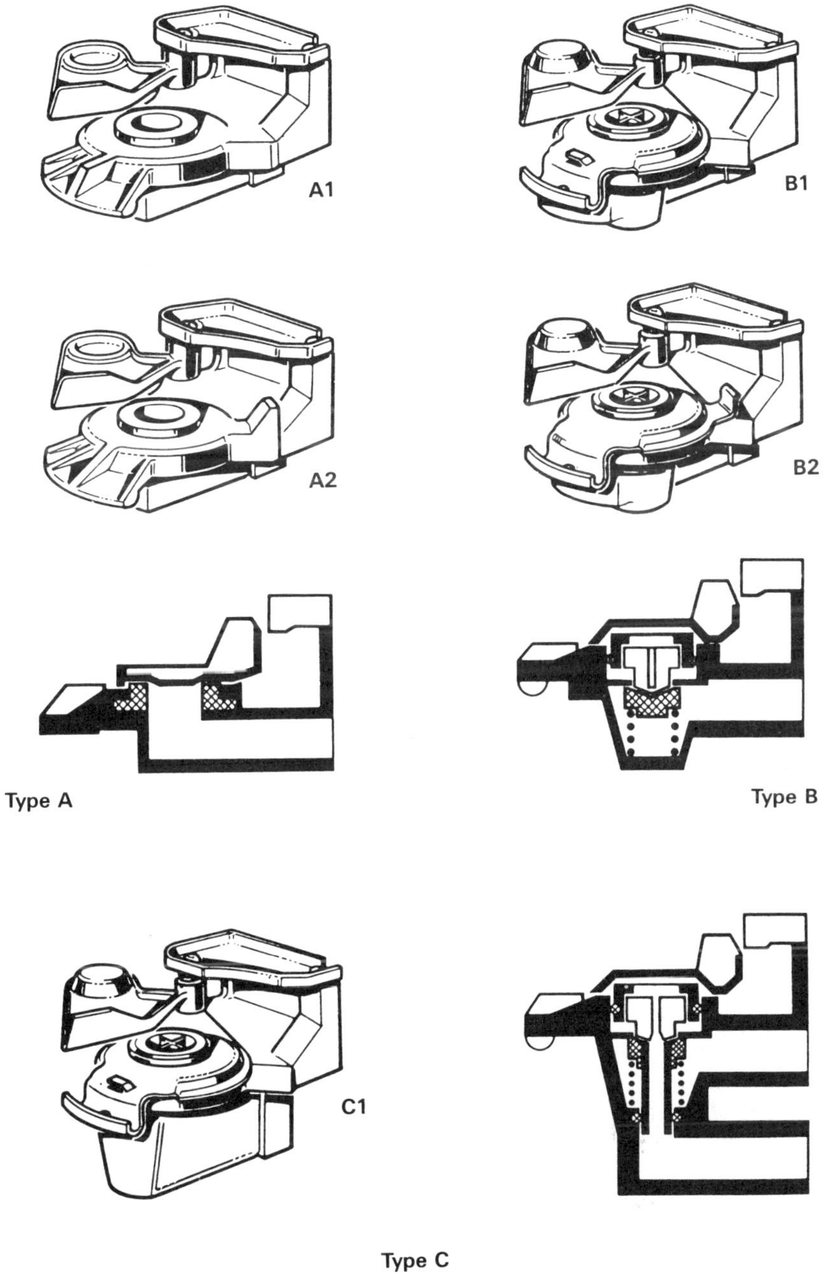

Figure 6.132 Coupling heads for two-line brake systems (FJ)

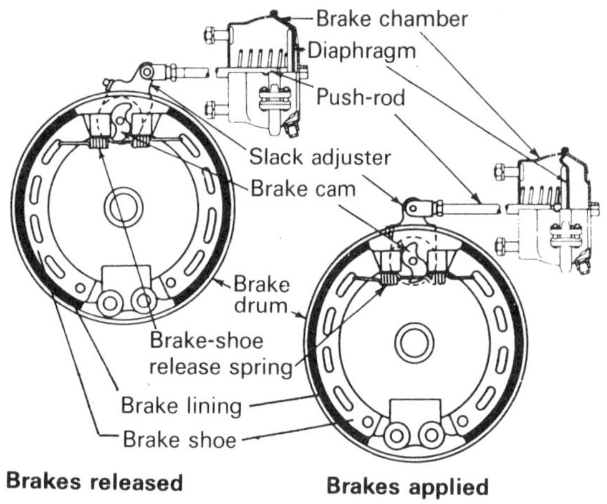

Figure 6.133 Slack adjuster (BW)

public service vehicles, heavy goods vehicles and railway rolling stock. One example is shown in Figure 6.135. To save repetition only one type is described. Figure 6.136 shows an exploded view of this component.

Working principle

The following explanation refers to both Figures 6.136 and 6.137 (on page 91). The brake chamber push-rod yoke is connected to the housing via a pin through a bush. The control unit assembly is fastened to the housing and contains the worm gear, meshed with a small worm gear on the shaft. A one-way clutch fits over the shaft , as do the needle bearing, thrust washer and coil spring. The external slot in the control unit is

Figure 6.135 Automatic slack adjuster (HX)

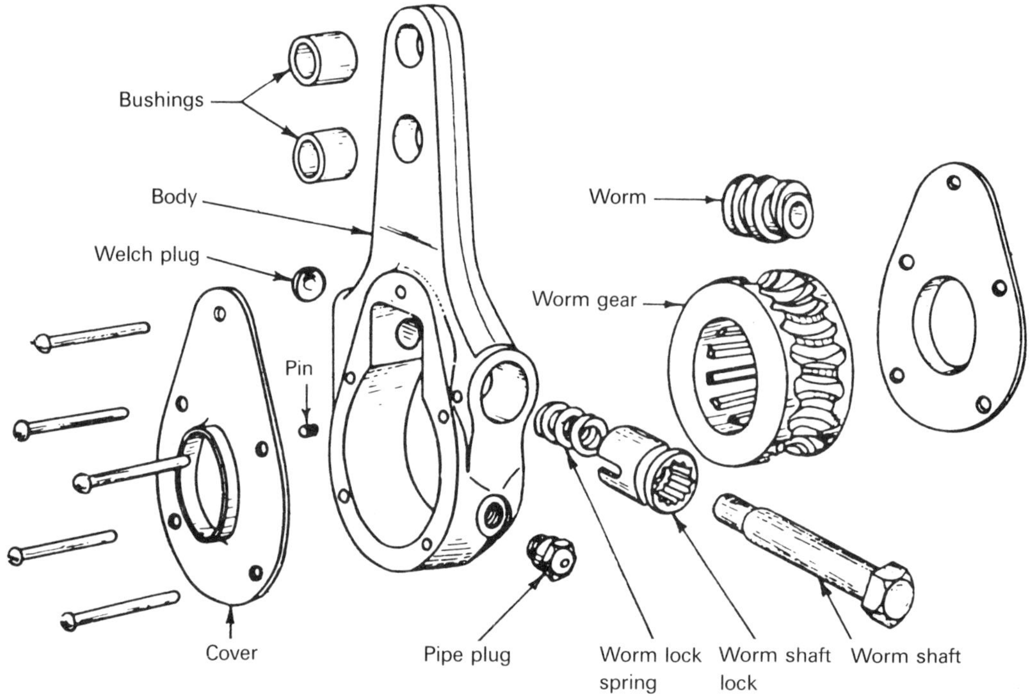

Figure 6.134 Component parts of a slack adjuster

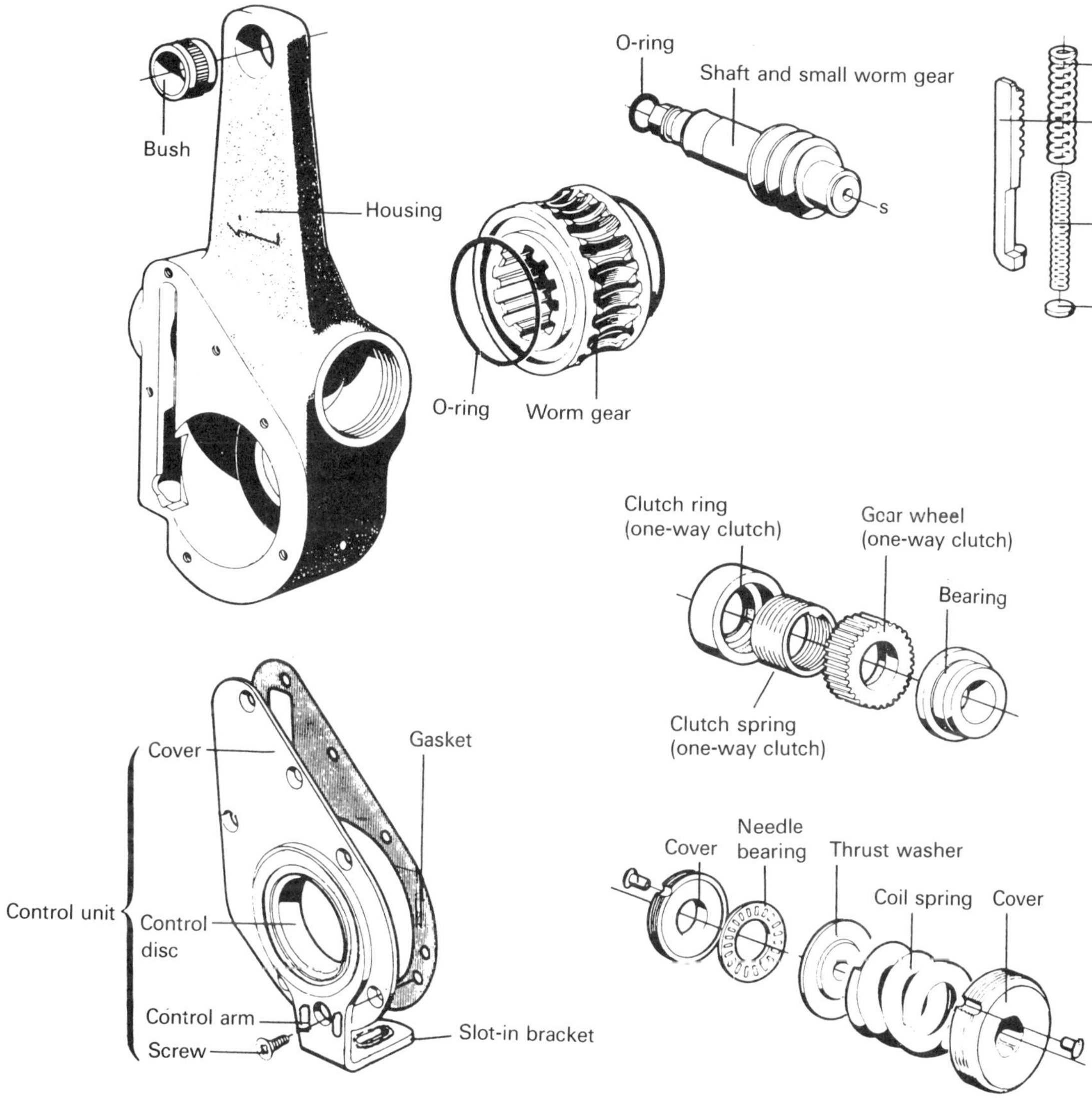

Figure 6.136 Automatic slack adjuster – component parts

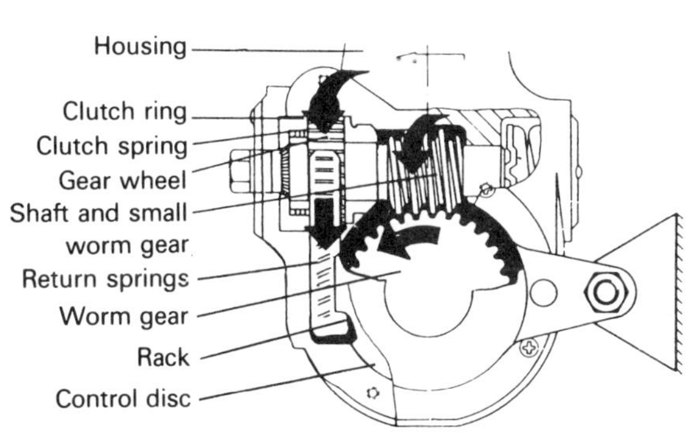

Figure 6.137 Automatic slack adjuster – operation

fastened to an adjacent component. The toe of the rack connects with the control unit and converts the rotary action of the control disc to straight-line movement of the rack.

Assuming that the clearance between brake lining and brake shoe is correct, when the brakes are applied the linings touch the drum as soon as the slack adjuster has rotated through the clearance angle between the bottom of the internal notch in the control disc to contact the toe of the rack. In these circumstances the adjustment is not altered.

If there is excessive clearance between the brake lining and the drum, after the slack adjuster has moved, as described above, the control disc pushes the rack upwards, causing the gear wheel to turn and disengage the one-way clutch. Simultaneously, the S-cam causes the brake shoes/linings to contact the brake drum.

The worm gear is moved along its shaft, compressing the coil spring and thereby disengaging the clutch. The control disc continues to push the rack upwards, and as a result turns the one-way clutch assembly. When the brakes are released the return springs keep the toe of the rack against the bottom of the slot in the control disc. The rack turns the one-way clutch assembly.

As the S-cam is released, the coil spring forces the small worm gear into contact with the clutch ring. The return springs do not exert sufficient force to turn the one-way clutch, when the cone clutch is engaged. Consequently, the toe of the rack is moved from the bottom to the top of the notch in the control disc, by the movement of that disc rotating clutch and worm to reduce shoe/drum clearance.

Manual adjustment, for use after brake relining, is achieved by turning the hexagon end of the worm shaft.

Air over hydraulics systems

'Air over hydraulics' is the term commonly used, where a foot valve supplies compressed air to an air actuator, which then applies the master cylinder, which in turn applies the wheel cylinders.

The design of the air actuator is similar to a brake chamber, but the main difference is that the end of the push-rod lacks thread, and has either a ball-shaped end, which is in contact with the piston of a master cylinder, or a disc which contacts the diaphragm in a neighbouring air actuator.

On modern vehicles, the foot valve, air chamber and master cylinder within an air over hydraulics system are all duplicated, to provide maximum safety in the event of a component failure.

Tandem air/hydraulic actuator (CD)

As shown in Figure 6.138, a dual foot valve, supplied with compressed air by two reservoirs, is connected to air chambers (A) and (B), which convert air pressure to mechanical force in order to operate a tandem master cylinder.

When the foot valve is applied, compressed air from its delivery ports, enters chambers (A) and (B) via ports (11) and (12) and acts upon diaphragms (a) and (b). This causes push-rod (c) to move to the right, towards the master cylinder, where it contacts piston (d), which pushes the recuperating valve (e) back to the vertical

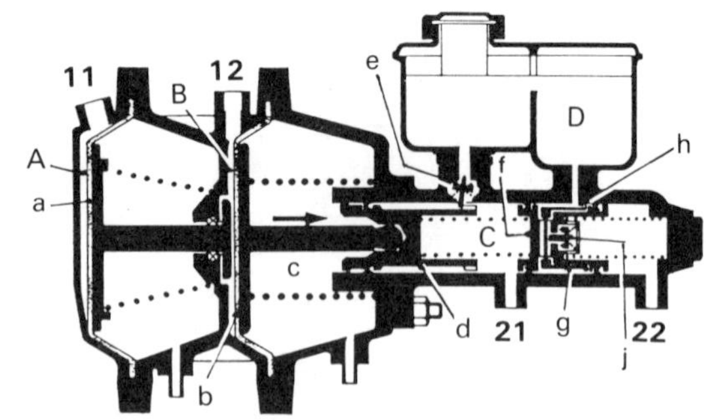

Figure 6.138 Tandem air/hydraulic actuator (CD)

(seated) position. The hydraulic pressure generated in chamber (C) acts upon piston (f), causing sleeve (g) to move away from stop (h) and allow recuperating valve (j) to be closed by its spring. Thus, hydraulic pressure leaves delivery ports (21) and (22) and is routed by bundy pipework to wheel cylinders, where their pistons operate the brake shoes mechanically.

When the foot valve treadle is released, air from air chambers (A) and (B) escapes to atmosphere via the exhaust port of the foot valve. Push rod (c) is returned to the 'off' position by the return springs, whilst pistons (d) and (f) also return to the 'off' position by the action of springs. Sleeve (g) returns to stop (h) and thereby lifts recuperating valve (j) off its seat, allowing brake fluid in reservoir (D) to top up the cylinder, as necessary. Likewise, as piston (d) returns to rest, it contacts and tilts the stem of recuperating valve (e), unseating the valve, and allowing brake fluid from the upper section of the reservoir to enter the cylinder.

In the event of a failure of air supply to either port (11) or (12), the surviving sound chamber will still activate the master cylinder.

Should either half of the master cylinder fail, the surviving sound circuit will operate *that section* of the braking system. In the defective part of the cylinder, the piston will 'bottom'; in the case of piston (d) against piston (f), and in the case of piston (f) against the cylinder end.

A device is fitted to some units to give warning of over-stroke. An electrical switch closes when a preset proportion of maximum push-rod travel is reached, and either illuminates a warning lamp or sounds a buzzer.

A float-type switch in the reservoir cap, similar to those used on cars, can be used to illuminate a warning lamp, to indicate low fluid level.

These hydraulic systems are produced in two different forms, one for use with orthodox brake fluid and the other for use with mineral oil. Naturally, only the correct fluid for the respective system should be used.

Trailer breakaway

If a trailer becomes detached from its towing vehicle, the brake connections between towing vehicle and trailer will break. It is, therefore, of the utmost importance that the trailer brakes are applied *automatically*. This is the role of the relay emergency valve.

Relay emergency valve (RE6) (CD)

Relay emergency valves are fitted to trailers to perform three important functions:

1 To apply the trailer brakes automatically if the trailer should break away from the drawing unit, or if the air pressure in the emergency line falls below approximately 30 lbf/in^2 (207 kN/m^2, 2 bar).
2 To speed up the application of the brakes by the use of a relay valve.
3 To speed up the release of the brakes by the use of the relay valve section.

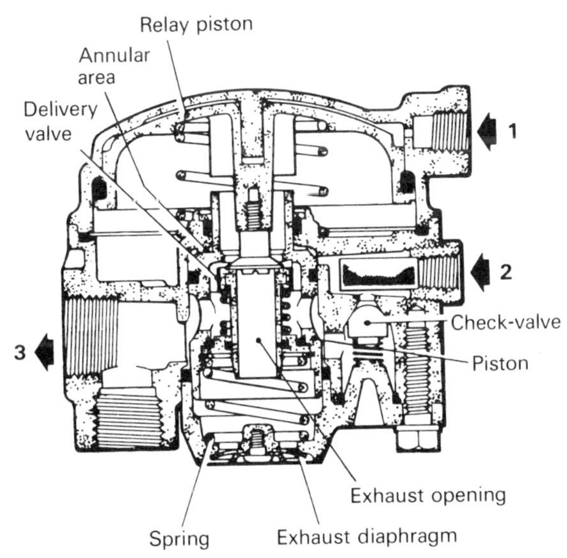

Figure 6.139 Relay emergency valve (RE6) (CD)

As shown in Figure 6.139, the main working components are a non-return valve, a relay piston and an emergency piston with an internal inlet and exhaust valve plunger.

Trailer brakes are operated, under normal circumstances, by compressed air that is supplied by the trailer reservoir, which is itself restocked by the emergency (supply) line from the drawing unit, via the relay emergency valve. To simplify the explanation, this is dealt with in stages.

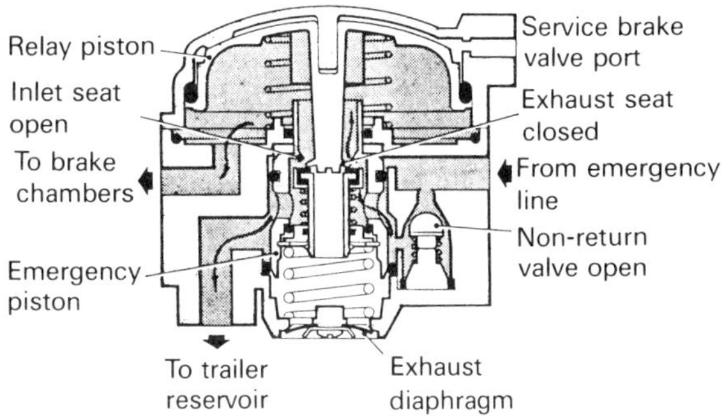

Figure 6.140 Relay emergency valve charging up to 60–65 lbf/in^2

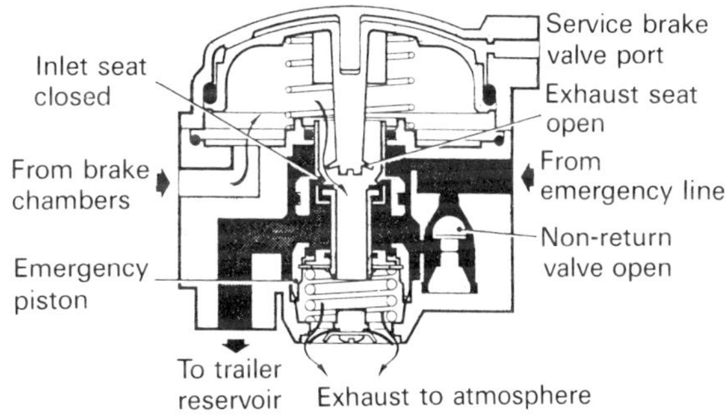

Figure 6.141 Relay emergency valve charging above 60–65 lbf/in^2

Charging: As shown in Figures 6.140 and 6.141, compressed air from the emergency (supply) line enters the emergency port and opens the non-return valve. It then passes through a drilling in the body, into the section containing the emergency piston. The compressed air passes through holes in the emergency piston, then continues around the inlet/exhaust valve stem, into the reservoir. Simultaneously, compressed air passes the open inlet valve, which is held off its seat by the exhaust seat stem of the relay piston, continuing on to the brake chambers to apply the brakes.

As the volume of air in the reservoir and brake chambers increases, the pressure also increases and this acts through a bleed hole in the casing, upon the emergency piston. This piston moves down the bore until the reservoir pressure reaches approximately 50 lbf/in^2 (345 kN/m^2, 3.5 bar) (the brake chamber pressure being the same), when the inlet seat on the emergency piston contacts the inlet/exhaust valve, cutting off further delivery of air to the brake chambers. When the reservoir pressure reaches approximately 60 lbf/in^2 (414 kN/m^2, 4 bar) the air pressure through the bleed hole, that is acting upon the emergency piston, causes the inlet/exhaust valve to be moved away from the exhaust seat, thus opening the exhaust passage through the inlet/exhaust valve stem. All the air from the brake chambers is then exhausted to atmosphere, through the diaphragm at the base of the unit, and the trailer can be driven away in safety.

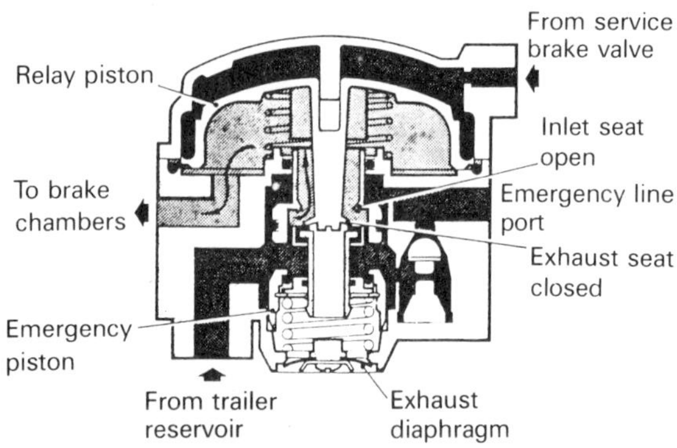

Figure 6.142 *Relay emergency valve applying position*

Applying: As shown in Figure 6.142, application of the service brake causes signal pressure to pass through the service line to the service brake port at the top of the relay emergency valve. This forces the relay piston downwards, and the exhaust seat contacts the inlet/exhaust valve, closing the exhaust passage in the valve stem. Further movement by the relay piston, moves the inlet/exhaust valve away from the inlet seat so that compressed air from the trailer reservoir can pass to the brake chambers and apply the brakes.

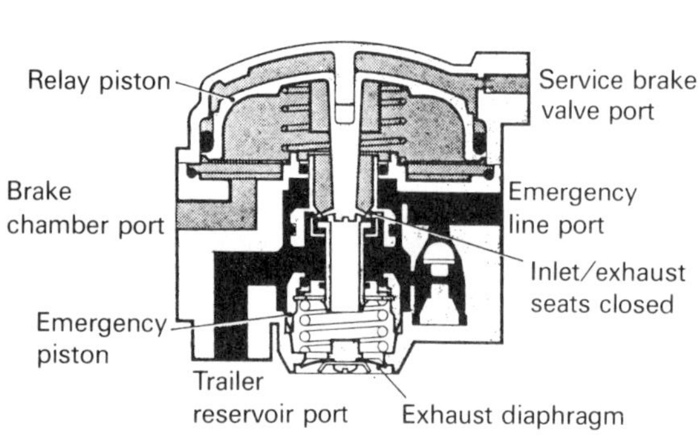

Figure 6.143 *Relay emergency valve balancing position*

Balancing: As shown in Figure 6.143, the pressure of the compressed air that is passed to the brake chambers is mirrored beneath the relay piston, which moves the piston upwards, permitting the inlet/exhaust valve to contact the inlet seat in the body. When this position is reached, with the inlet and exhaust seats closed, the valve is said to be in a balanced condition. The pressure above and below the piston is equal. However, if the driver either increases or decreases pressure on the foot valve treadle, the relay piston will respond and rectify the situation until balance is restored, but with the new pressures involved.

If more pressure is applied the piston would unseat the inlet/exhaust valve from the inlet seat allowing more compressed air to flow from the reservoir to the chambers, until balance is restored. In the case of reducing pressure, when pressure is lost from above the piston, the higher pressure below lifts the piston and the exhaust seat upwards, allowing some air to escape from the chambers to the atmosphere, before restoring balance at a lower pressure.

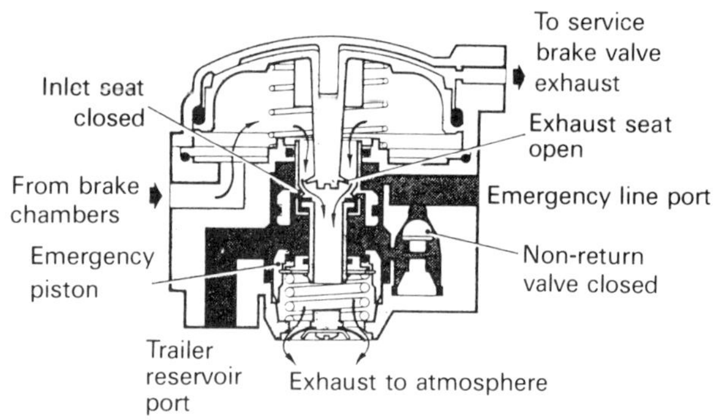

Figure 6.144 *Relay emergency valve releasing position*

Releasing: As shown in Figure 6.144, when the footbrake treadle is completely released, all the pressure above the relay piston is released through the brake valve, whilst the pressure beneath the relay piston moves it up to the highest position. This opens the exhaust passage, allowing the air from the chambers to escape to atmosphere and the brakes to release.

Emergency: As shown in Figure 6.145, if the pressure in the emergency line drops to below 30 lbf/in^2 (207 kN/m^2, 2 bar), the return spring will push the emergency piston upwards. This will close the inlet/exhaust valve against the exhaust seat and then draw the inlet seat away from the inlet/exhaust valve,

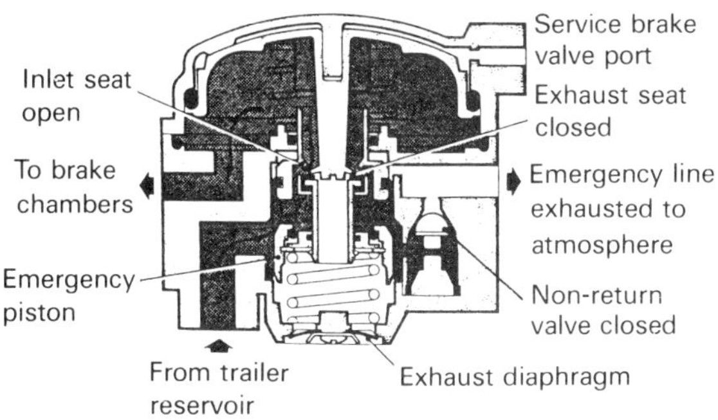

Figure 6.145 Relay emergency valve emergency position

allowing air from the trailer reservoir to reach the brake chambers. The non-return valve fitted to this unit prevents loss of air through a leak in the emergency line.

Although RE6 emergency valve is in common use, the older type (RE4), with the exhaust port at the top of the unit, is still to be found on older trailers. The range of modern relay emergency valves also includes the diaphragm type (see Figure 6.146) and the relay emergency valve with dynamic predominance (see Figure 6.148).

Relay emergency valve – diaphragm-type (WO)

This valve, which is fitted to a trailer, serves three purposes: it speeds up the application of the trailer brakes; it speeds up the release of the trailer brakes; and it applies the trailer brakes automatically if there is a fracture in the red emergency (supply) line, between tractor and trailer.

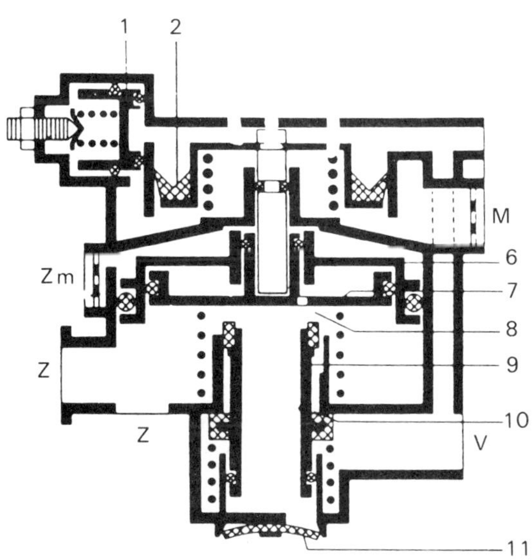

Figure 6.146 Relay emergency valve – diaphragm-type (WO)

As shown in Figure 6.146, compressed air from the tractor reservoir passes through the red emergency (supply) line and enters port (M), before passing seal (2), and then leaving the relay emergency valve via port (V), en route to the trailer reservoir.

When the air pressure in the trailer reservoir reaches 85 lbf/in^2 (586 kN/m^2, 6 bar), pressure balance valve (1) opens. When there is a pressure drop in the tractor system, it allows air to flow back to the tractor reservoir, closing between 74–77 lbf/in^2 (510–530 kN/m^2, 5.1–5.3 bar). If such compensation of pressure was not possible, the trailer would be subjected to automatic application of the trailer brakes every time there was a drop in pressure. Naturally, if the red emergency (supply) line fractured, the trailer brakes would be automatically applied.

Service brake application: When the foot valve treadle is applied, compressed air is passed through the yellow service (control) line to port (Zm). It will then act upon the upper face of piston (6), moving it down, together with piston (7), which closes exhaust valve (8) and opens inlet valve (10). This allows air from the trailer reservoir to flow into port (V) and to leave again via port (Z), en route to the brake actuators.

When the air pressure below pistons (6) and (7) reaches the signal pressure above, the pistons move up, as does hollow stem (double-seated valve) (9), allowing the inlet valve (10) to be closed by its return spring. Thus, the exhaust route is closed and the inlet valve is closed, leaving signalled air pressure in the brake chamber and keeping the brakes applied. The valve is then in the lapped or balanced condition. It will remain thus, until the driver alters the force that he is applying to the foot-treadle.

When the driver removes his foot from the treadle, signal pressure to port (Zm) will cease, and pistons (6) and (7) will move to their highest position. Exhaust valve (8) will open and compressed air will exhaust from the trailer brake chambers through the hollow stem (double-seated valve) (9) and the exhaust (flap) check valve (11).

Trailer-release valve (WO)

The relay emergency valve shown in Figure 6.146 can be fitted with a trailer-release valve (see Figure 6.147), to enable a trailer to be moved after it has been disconnected from its tractor. Under normal circumstances the trailer brakes would be fully applied as a result of the action of the relay emergency valve.

To obtain the brake release, the knob and stem (13)

are pulled out to the stop, which seals off the airway between ports (12) and (14), opening the airway from port (15) to port (14). Therefore, trailer reservoir pressure at port (14) enters the relay emergency valve at port (M) in Figure 6.146, which moves the relay valve down into the normal driving position, and compressed air is exhausted from the trailer brake chambers, via the relay-emergency valve, thus releasing the trailer brakes.

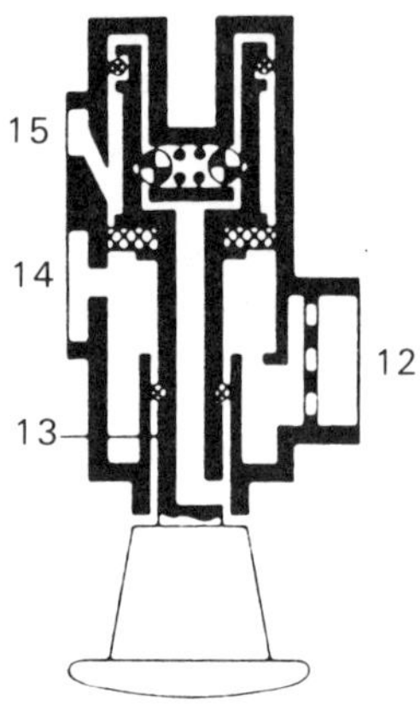

Figure 6.147 Trailer-release valve (WO)

When the tractor and trailer are reconnected, if the release-valve knob has not been pushed back, air from the relay emergency valve, entering via port (12) will push stem (13) inwards, returning this valve to the running condition, with ports (12) and (14) re-connected.

A vent hole in stem (13) provides 'cushioning' when the knob is pushed in, and it also prevents a vacuum forming when the knob is pulled out.

Relay-emergency valve with dynamic predominance (WO)

This valve is fitted into the braking system of a semi-trailer and is controlled by the air pressure from the tractor.

It ensures:

1 that the application and the release of the trailer brakes are in relation to the actuation of the tractor brakes, by means of a special device suitable for a long, multi-axle semi-trailer, with a fifth port for tri-axles;
2 automatic or emergency braking, if a partial or total loss occurs in the red emergency (supply) line.

As shown in Figure 6.148, compressed air from the red emergency (supply) line from the tractor, enters port (1). It passes through seal (b), to supply the trailer reservoir via port (1-2). At the same time piston (c), together with valve (e), moves downwards as a result of the air pressure above. Exhaust valve (a) opens and allows the air pressure in the brake chambers, connected to the ports (2), to exhaust to atmosphere via check valve (3).

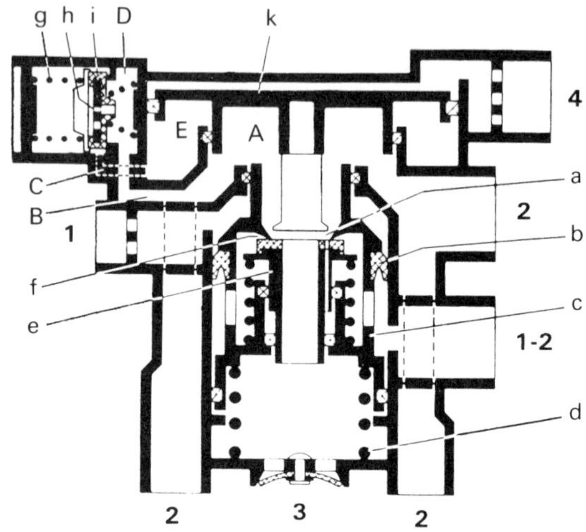

Figure 6.148 Relay-emergency valve (with dynamic predominance) (WO)

Service brake applied: A signal pressure from the foot valve, on the tractor, enters port (4) and acts upon piston (k) which moves downwards, closing exhaust valve (a) and opening inlet valve (f). This connects the trailer reservoir to the trailer brake chambers. Simultaneously, air flows through passage (B) into chamber (D), through calibrating orifice (h) and then via duct (C) into chamber (E).

Owing to the difference in the cross-sectional area, above and below piston (k), the pressure in chamber (A), controlled by pressure at port (4), is achieved rapidly. The pressure builds up progressively in chamber (E) and because of this piston (k) reaches a state of balance to give an outlet pressure at port (2), equal to the signal pressure at port (4). At this stage valve (e) moves up and shuts off the supply.

A predominance device, pressurised through the calibrating orifice (h) is mounted on valve (i), and held by spring (g), to avoid a dangerous pressure build-up occurring if the orifice became blocked.

When the service brake treadle is released, pressure falls at port (4) and piston (k) rises, opening exhaust valve (a). Air pressure in the trailer brake chambers is exhausted to atmosphere via exhaust check valve (3). Air pressure in chamber (E) is released through duct

(C), passage (B) and chamber (A), exhausting to atmosphere via exhaust check valve (3).

If there is a fracture in, or a disconnection of, the red emergency (supply) line, port (1) will exhaust to atmosphere. Thus, piston (c) will not be held down by air pressure acting upon its top surface, and consequently it will be forced upwards by the trailer reservoir pressure, and the force of the emergency spring (d). This closes exhaust valve (a), opens inlet valve (f) and allows the full air pressure, within the trailer reservoir, to pass to the trailer brake chambers and fully apply the trailer brakes.

On a modern vehicle, if the yellow service (control) line from the tractor to the trailer fractures, this would cause the trailer control valve (or dump valve), located on the tractor, to exhaust the red emergency (supply) line. This would cause the trailer brakes to be fully applied automatically, as a result of the same action by the relay emergency valve, as previously described.

Front-axle ratio valve (BW)

This valve (shown in Figure 6.149) is fitted to American heavy goods vehicles (see Figure 7.13 on page 115). Its role is to limit the amount of air pressure delivered to front brake chambers, as a result of normal service brake applications. When the foot valve is applied, compressed air enters the top port (supply) and acts upon the top face of the inner and outer pistons.

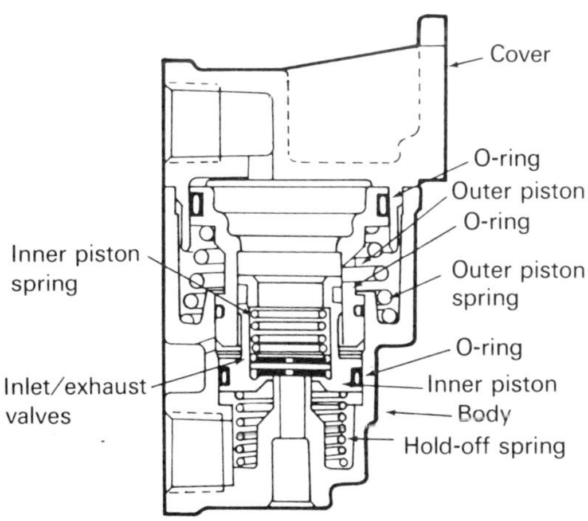

Figure 6.149 Front-axle ratio valve (BW)

For applications where the input pressure is less than 40 lbf/in^2 (276 kN/m^2, 3 bar), that pressure is not sufficient to depress the outer piston which has a strong return spring, and thus only the inner piston is moved downwards, closing the exhaust port. Compressed air then flows out of the delivery port to the brake chambers. As pressure builds up in those chambers, it also builds up beneath the inner piston in this valve. When the delivery pressure reaches half of the value of the supply pressure, this will cause the inner piston to lift upwards, closing the inlet valve, but leaving the exhaust valve closed. At this stage the valve is in the lapped, or balanced, condition.

Note: With this valve, an input pressure of 30 lbf/in^2 (207 kN/m^2, 2 bar) at the supply port, would result in 15 lbf/in^2 (103 kN/m^2, 1 bar) output pressure at the delivery port).

When there is a greater pressure than 40 lbf/in^2 (276 kN/m^2, 3 bar) to the supply port, the pressure at the delivery port does not drop to the same degree as before, and this reduction ratio gets progressively smaller, until at 60 lbf/in^2 (413 kN/m^2, 4 bar) the supply and delivery pressures are equal. This is due to the higher pressure overcoming the outer piston return spring.

When the foot valve is released, supply ceases to this front-axle ratio valve, and both pistons return to TDC, by spring pressure, causing the inlet/exhaust valve to unseat from the exhaust stem, thus providing a direct route to atmosphere from the front brake chambers.

MV-2 control module (BW)

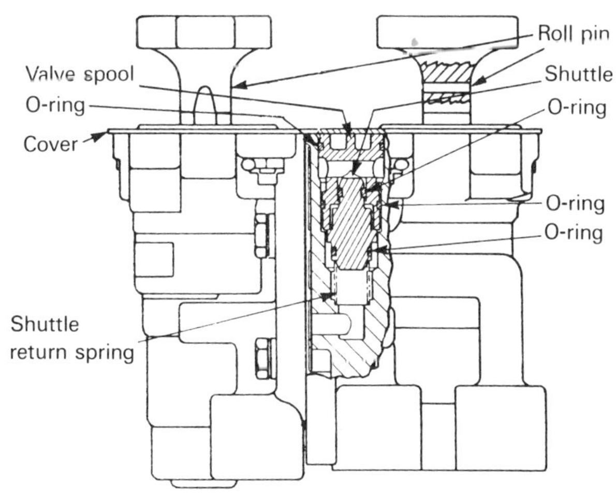

Figure 6.150 MV-2 control module (BW)

This component (see Figure 6.150) is fitted to some vehicles that are produced in the USA (see Figure 7.13 on page 115). It provides the following options:

tractor (secondary) spring-brake control,
trailer (either orthodox or spring) brake control,

parking brake control of the whole combination,
trailer brake (only) control (non-graduable),
trailer system charging whilst the tractor brakes are applied,
selection of supply reservoirs.

It is a small unit, suitable for mounting on the control panel (dashboard) in the tractor cab. There are two driver operated, push-pull buttons (one red and one yellow), used to operate two integral valves. However, if there is a severe loss of pressure within the system, these over-ride the driver's choice and apply the brakes automatically. The red button is marked 'Trailer air supply. Pull to evacuate. Push to supply. Not for parking.' The yellow button is marked 'Parking brake. Pull to apply. Push to release.'

Incorporated into this unit is a spring-loaded, dual-circuit supply valve, which accepts air on behalf of both control valves from number one reservoir, unless the pressure in that reservoir falls below 30 lbf/in^2 (207 kN/m^2, 2 bar), when the dual-circuit supply valve will shuttle and number two reservoir will become the source of supply.

The trailer air supply button (red) delivers air to the red emergency (supply) line to the trailer, but it will also pop out automatically, shutting off the trailer supply, if pressure drops to approximately 40 lbf/in^2 (276 kN/m^2, 3 bar).

The parking brake button (yellow) controls the spring brakes on the tractor, and at the same time causes the trailer air-supply valve to exhaust, thereby additionally applying the trailer parking brakes. The trailer brakes can then be released by pushing in the red button.

System at zero pressure: Both buttons will be out. Upon charging the system with compressed air to 65 lbf/in^2 (448 kN/m^2, 4.5 bar), the red button (trailer supply) can be pushed in and will remain in, allowing the trailer system to be charged and the trailer brakes to be released. The yellow button can then be pushed in, and it will supply air to release the spring brakes fitted to the tractor.

Normal driving: With tractor and trailer systems at safe working air pressure, all brakes are in the 'off' condition, and both the yellow and red buttons are retracted.

Operation of trailer emergency brakes: The red button is pulled out and this will apply the trailer brakes. If spring brakes are used this will be by exhausting air. With orthodox brakes, air will be supplied from the trailer reservoir via a relay emergency valve.

Combination park: By pulling the yellow (parking brake) button out, this will cause the red (trailer air supply) button to pop out, as the air evacuates from the tractor spring brakes. Thus the brakes on both tractor and trailer are applied.

Charging trailer system: When the combination is parked, if it is necessary to recharge the trailer system, whilst leaving the spring brakes applied on the tractor, the red button should be pushed in.

Automatic brake application: If during normal running, air pressure supplying the push-pull valves is reduced to approximately 40 lbf/in^2 (276 kN/m^2, 3 bar), the red button (trailer air supply) will pop out and the emergency brakes on the trailer will be applied. If that red button is then held in, and the air pressure drops to approximately 30 lbf/in^2 (207 kN/m^2, 2 bar), a tripper piston within this unit will over-ride the manual setting and again operate the trailer brakes. The yellow button (parking brakes) will pop out at approximately 20–30 lbf/in^2 (138–207 kN/m^2, 1.5–2.0 bar), rendering the complete combination immovable, until some corrective action is taken.

SR-4 spring-brake control valve (BW)

This unit is fitted to some trailers that are manufactured in the USA (see Figure 7.14 on page 116). Its function is to allow spring brakes to be applied both under driver control and automatically.

As shown in Figure 6.151, compressed air from the red emergency (supply) line from the tractor supplies compressed air to the trailer supply port of this valve. This air pressure opens the spring-brake inlet valve, against its return spring. The pressurised air also travels to the chamber beneath the pressure-protection piston, and when the pressure builds up to approximately 55 lbf/in^2 (380 kN/m^2, 4 bar), this piston is moved up against the force of its return spring, opening pressure-protection inlet valve.

Air pressure now flows past check-valve (A), and into the rear service reservoir. It also flows past check-valve (B), through the spring-brake inlet valve and into the spring-brake chambers, via a cavity immediately behind the spring-brake inlet/exhaust valve. At the same time air enters pressure-protection valve PR-3. As the pressure in the pressure-protection valve PR-3 reaches approximately 60 lbf/in^2 (413 kN/m^2, 4 bar) its inlet valve opens, allowing air to flow past the check-valve (in pressure-protection valve PR-3) and into the front-service reservoir. Check-valve (C) prevents air entering the front service reservoir from the upstream side of the PR3. The spring brakes are now fully released.

Park: The trailer supply line is exhausted by the operation of a cab control. Air pressure is reduced to zero at the control piston and at the internal pressure-protection piston, causing both the pressure-protection valve and the spring-brake inlet valve to close, exhausting the spring-brake chambers and causing the springs to apply the foundation brakes.

Service brake failure: If the air pressure in both the tractor and trailer systems drop until the internal pressure-protection valve shuts and maintains 50–60 lbf/in^2 (344–413 kN/m^2, 3.5–4.0 bar), this will keep the spring brakes released. However, a low-pressure signal will be provided in the cab to warn the driver.

If the service failure affects the rear service reservoir, air flowing to the pressure-protection valve will escape via the rear-service reservoir and not resupply the front reservoir. Nevertheless, this front reservoir will retain full pressure because of the check valve in PR-3 and check valve (B). This reserve of compressed air will be sufficient to provide at least one spring-brake release, after an application by the driver.

Service applications will still be available until the compressed air is exhausted from the front reservoir.

With the failure at the front-service reservoir, pressure-protection valve PR-3 will close, and air passing the internal pressure-protection valve will supply the rear reservoir, to be available to release the spring brakes, through check valve (B). The service brake will not be available for use.

Emergency application, when service brake has failed: After a failure of the service brake system, the remaining 55 lbf/in^2 (380 kN/m^2, 4 bar) in the trailer supply line can be exhausted either manually or automatically (by trailer supply valve or parking control valve). The spring brakes are then applied, exhausting through the control piston. Sufficient pressure will remain in the trailer system to enable the spring brakes to be released at least once.

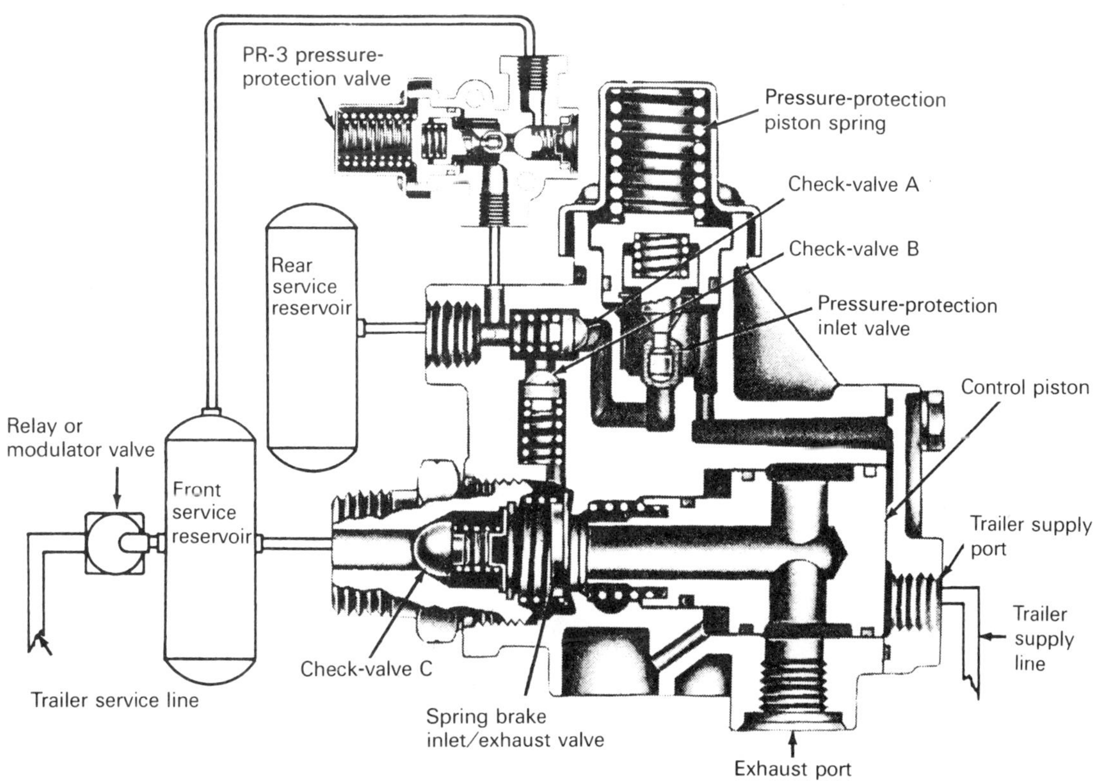

Figure 6.151 SR-4 spring-brake control valve (BW)

7 Piping diagrams, built up in stages

Piping diagrams

The piping diagrams found on pages 101 to 116 show a variety of braking systems which illustrate how many of the components described in this publication can be incorporated into modern layouts. However, it should be noted that these are only a few of the layouts that can be designed and installed in vehicles. Variations can range from the substitution of one component for another to the complete rearrangement of pipework and/or components.

Common key to Figures 7.1 to 7.14 inclusive

1 = Alcohol evaporator
2 = Alcohol injector
3 = Automatic drain valve
4 = Buzzer
5 = Differential-protection valve
6 = Double-check valve
7 = Drain cock
8 = DV3 auto-drain valve
9 = Delta dual foot valve
10 = E1 foot valve
11 = E6 foot valve
12 = Governor valve
13 = Handbrake valve
16 = L4 spring-brake hand valve
17 = Low-pressure indicator
18 = Multi-circuit protection valve
19 = Pressure-protection valve
20 = Pressure-reducing valve
21 = Pressure-regulating valve
22 = Quick-release valve
23 = Relay valve
24 = Relay emergency valve
25 = S4 hand valve
26 = S6 hand valve
27 = Safety valve
28 = Unloader valve
29 = Check valve
30 = Single-circuit protection valve
31 = Slack adjuster
32 = Stop-light switch
33 = Supply dump valve
34 = TC2 hand valve
35 = Test connector
36 = Trailer control valve (plus differential and supply dump)
37 = Trailer control valve (plus predominance and supply dump)
38 = Two-way valve
39 = Variable-load valve
40 = Trailer control valve (basic)

Single-line system (Figure 7.1)

A compressor supplies air to the reservoir via an unloader valve. A safety valve is fitted as a safeguard against excessive pressure build up. The foot valve is supplied with compressed air from the reservoir, and when the treadle is depressed, controlled air pressure is supplied to the brake chamber. Warning of low air pressure is visually shown on the pressure gauge and the buzzer gives audible warning.

Dual system – truck (Figure 7.2)

A compressor supplies air to three reservoirs via an unloader valve. There are two separate service systems, with each half of a dual foot valve being independently supplied with air from the relevant section of a divided reservoir. Air assistance is supplied by the third reservoir when the hand brake lever is applied. When this lever is set in the park position, the compressed air is released and the lever is held by a mechanical device such as a pawl and ratchet.

Three-line tractor system (Figure 7.3)

A governor valve is fitted to this vehicle, instead of an unloader valve, to control the working pressure of the air in the reservoirs. One of the two reservoirs supplies air to the single foot valve, which, when the treadle is applied, supplies air to the service port on the double-diaphragm brake chambers on the tractor. Simultaneously, the foot valve supplies air to the service port of each double-diaphram brake chamber on the trailer, via the relay valve on the tractor, the yellow (control) service flexible hose and the relay emergency valve on the trailer.

A secondary circuit reservoir supplies air to the hand valve, which when applied, supplies air to the secondary port of the double-diaphragm brake

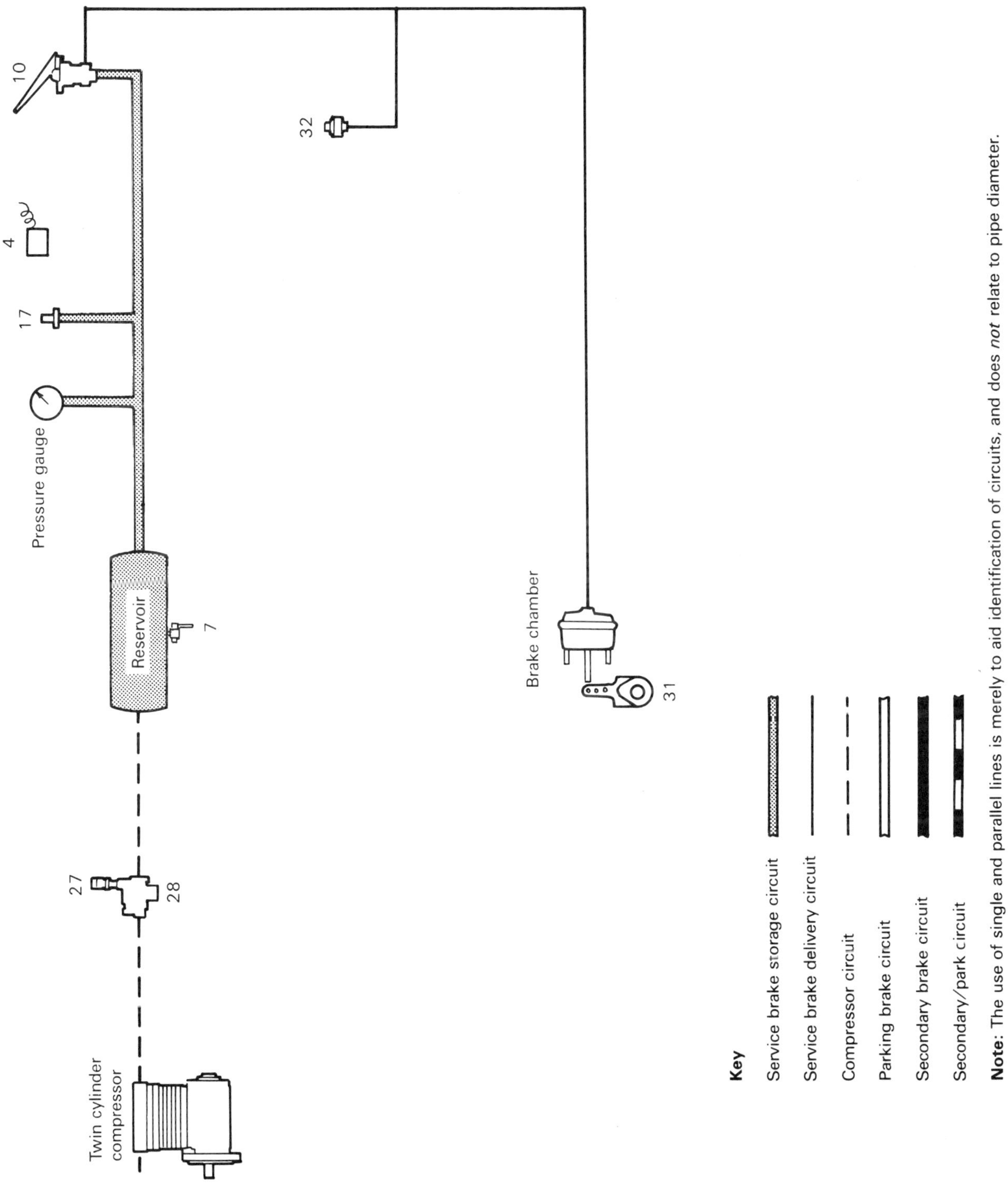

Figure 7.1 Single-line system

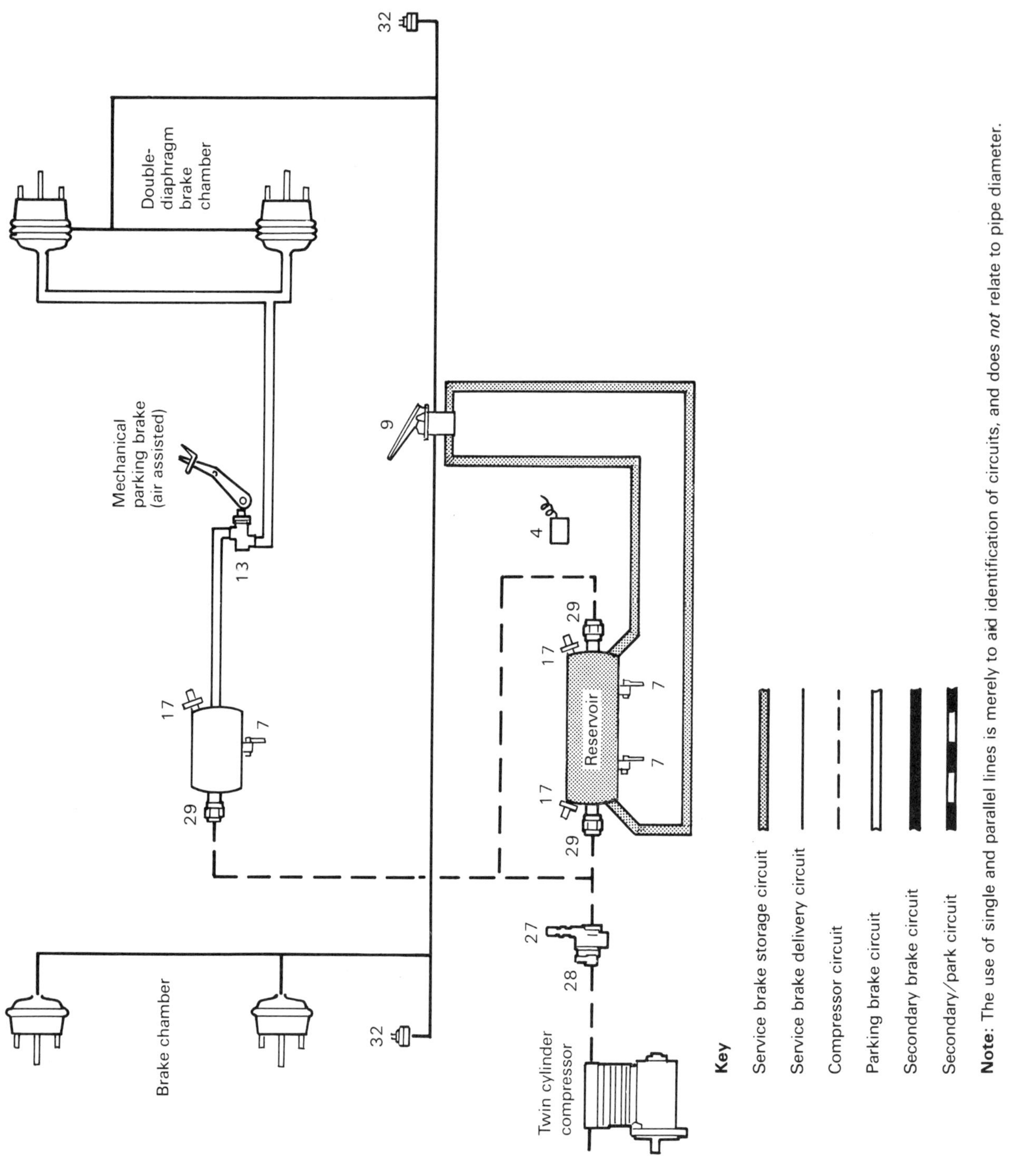

Figure 7.2 Dual system – truck

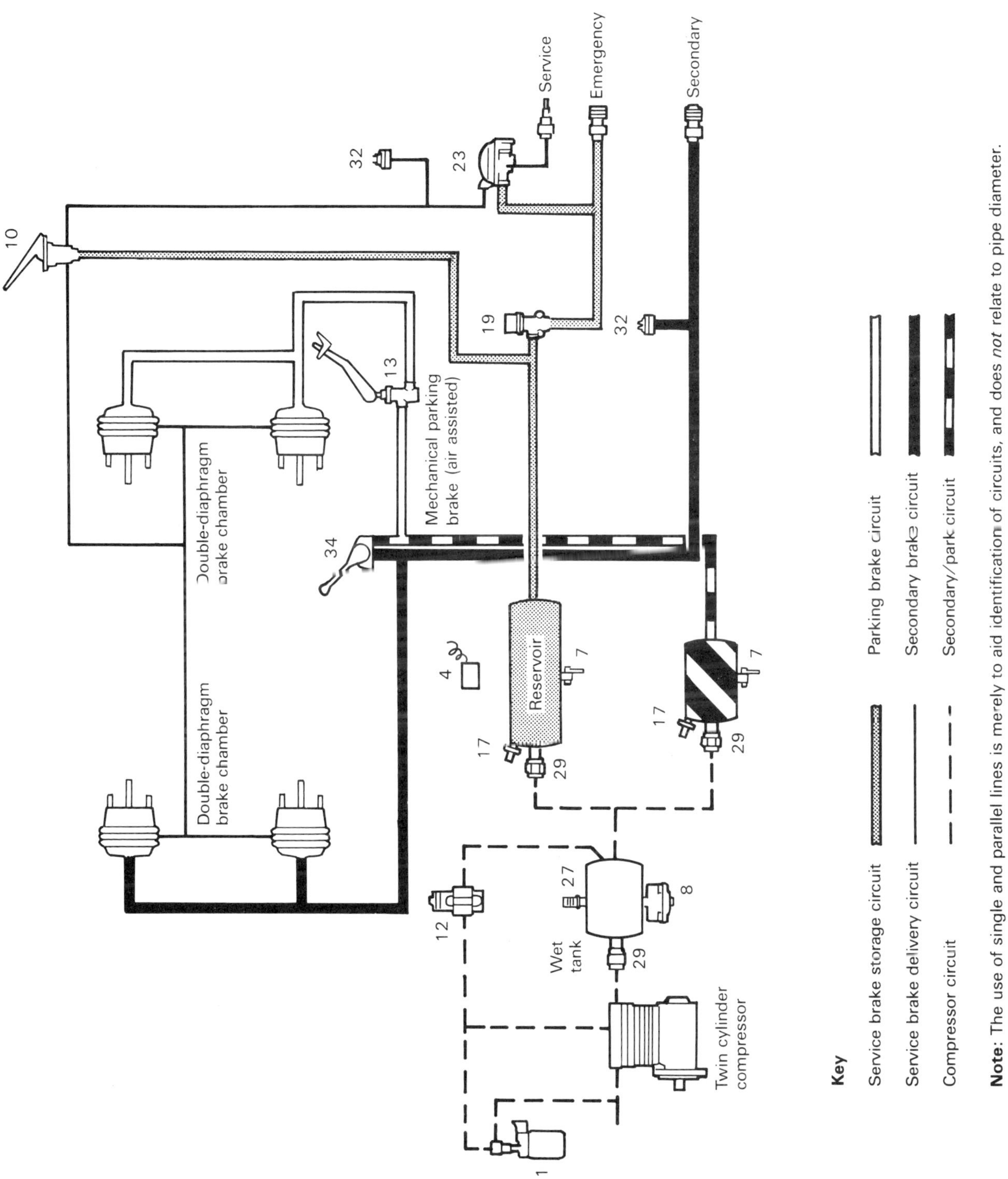

Figure 7.3 Three-line tractor system

chambers on the front axle of the tractor and to the secondary port of the brake chambers on the trailer. Note that the drive axle on this vehicle is *not* braked by the secondary brakes.

Owing to the potential danger of air leaks, a mechanical handbrake is a legal requirement, where this type of hand valve is fitted. The handbrake fitted to this vehicle is air-assisted, merely to aid the driver, but the air is exhausted when it is set in the park position. The brakes are held on mechanically, typically by a pawl and ratchet. A pressure-protection valve is fitted in the pipeline to the red emergency (supply) line to the trailer, to protect the tractor brakes in the event of a trailer break-away.

Three-line trailer system (Figure 7.4)

There is a separate air reservoir on the trailer, which is supplied with air from a reservoir on the tractor via the red emergency (supply) flexible hose and the relay emergency valve. Also connected to the relay emergency valve is a pipe connection that is linked to the foot valve on the tractor via the yellow service (control) flexible hose. In addition, other flexible hoses connect the relay emergency valve with the service port of each double-diaphragm brake chamber on the trailer.

When the red emergency (supply) line between the tractor and trailer is disconnected or pressure in that line drops to a low level, all of the trailer brakes are fully applied by air from the trailer reservoir, supplied through the relay emergency valve.

When the foot valve on the tractor is applied, the application of the service brakes on the trailer is speeded up, via the relay emergency valve. Similarly, the release is speeded up as air exhausts to atmosphere through the exhaust port of this valve.

The hand valve on the tractor is connected to the secondary ports of the trailer brake chambers via a blue secondary (auxillary) flexible hose and a quick-release valve, but it does not connect with the relay emergency valve.

Truck – spring brakes (Figure 7.5)

Application of the foot-treadle supplies compressed air to the service port of each (combined) spring-brake chamber.

The secondary system and parking brake are controlled by the hand valve. When the lever is moved towards secondary/park, air pressure in the secondary line is exhausted from the exhaust port of the hand valve, relay valve and quick-release valve, whereupon the spring within each chamber applies the brakes.

A differential-protection valve is fitted to prevent both sections of the brake, i.e. air and spring, being applied together, causing damage to components by the compounding of forces.

A conventional stop-light switch is fitted in the service line, but because the secondary/park brakes are applied by reducing pressure, the stop lights in this circuit are operated by a low-pressure switch.

Tractor – spring brakes (1) (Figure 7.6)

When the foot-treadle is applied, compressed air is supplied to the service diaphragm in each brake chamber.

Compressed air is supplied from the secondary/park reservoir to an L-gated hand valve (L4) and to two relay valves. One of the relay valves is in the secondary line to the trailer and the other is in the secondary line to the tractor. When the hand lever is in the 'off' position air is fed to the relay valve in the spring-brake line and, therefore, the spring brakes on the tractor are held off. At the same time the secondary line to the trailer is exhausted, releasing the trailer brakes.

As the hand lever is moved towards the secondary position, air in the spring line is reduced, and additionally air is introduced into the secondary line to the trailer.

With the hand lever in the park position, the tractor is held stationary by the spring brakes, but the trailer brakes are off.

A test point is shown at either side of the variable-load valve. Both service and secondary circuits, in this case, are fitted with conventional stop-light switches.

Tractor – spring brakes (2) (Figure 7.7)

Each half of the foot valve is supplied from a separate reservoir. Air supplied by one half of the dual foot valve acts on the service diaphragm, in the service chamber, on each of the spring brakes on the front axle, whereas the other half acts on similar diaphragms on the rear axle. This system is known as a *front/rear split*. A further safety feature is that each half of the foot valve acts on the double-check valve of an R62DC relay valve, in the service (control) line to the trailer.

Secondary braking, on the drawing unit, is by spring brakes on both axles, and on the trailer by upright air. Parking is by spring brakes on the tractor only and not on the trailer.

Air/hydraulic tractor (Figure 7.8)

This system generates and stores compressed air in the conventional manner. It also incorporates all of the

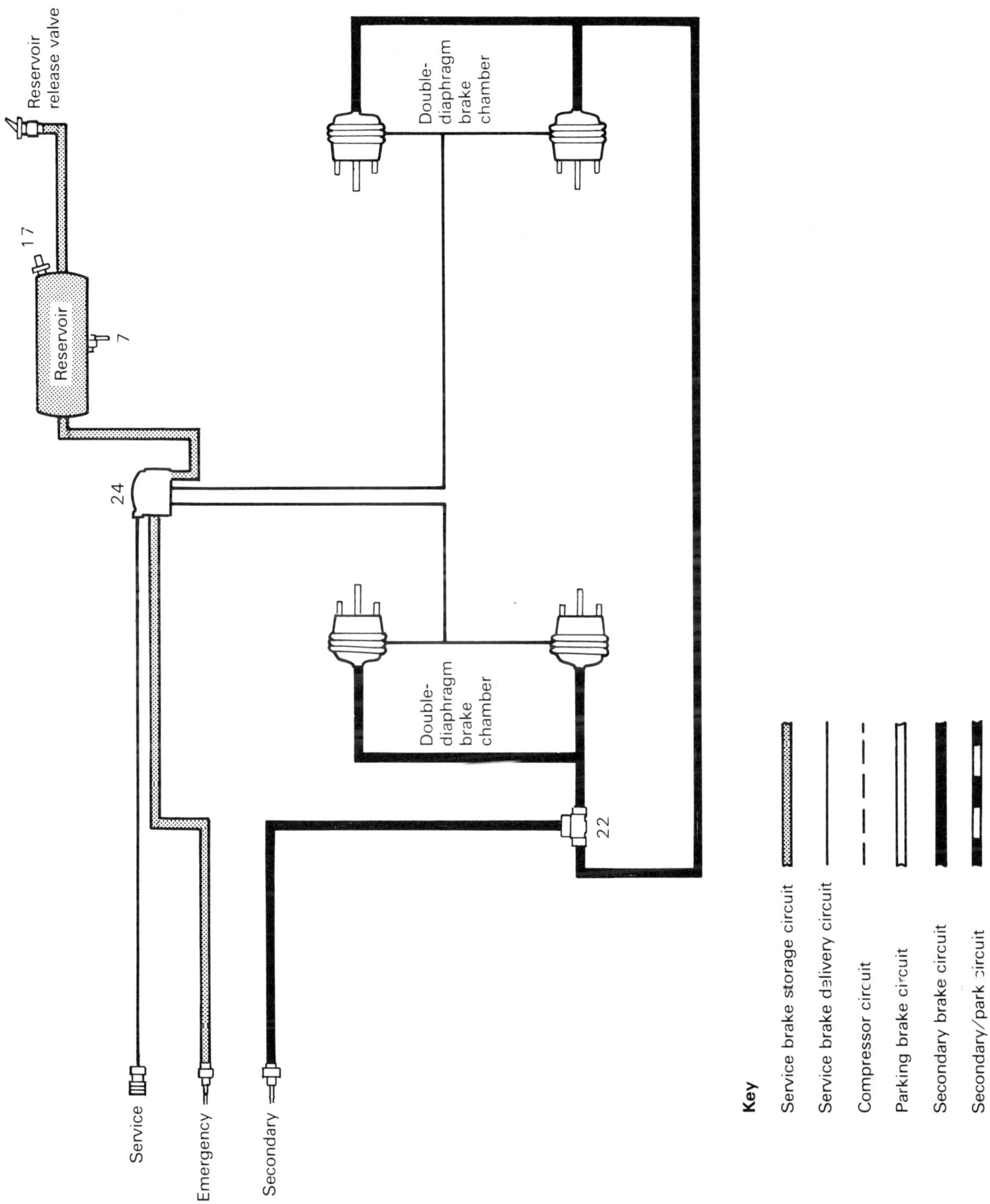

Note: The use of single and parallel lines is merely to aid identification of circuits, and does *not* relate to pipe diameter.

Figure 7.4 Three-line trailer system

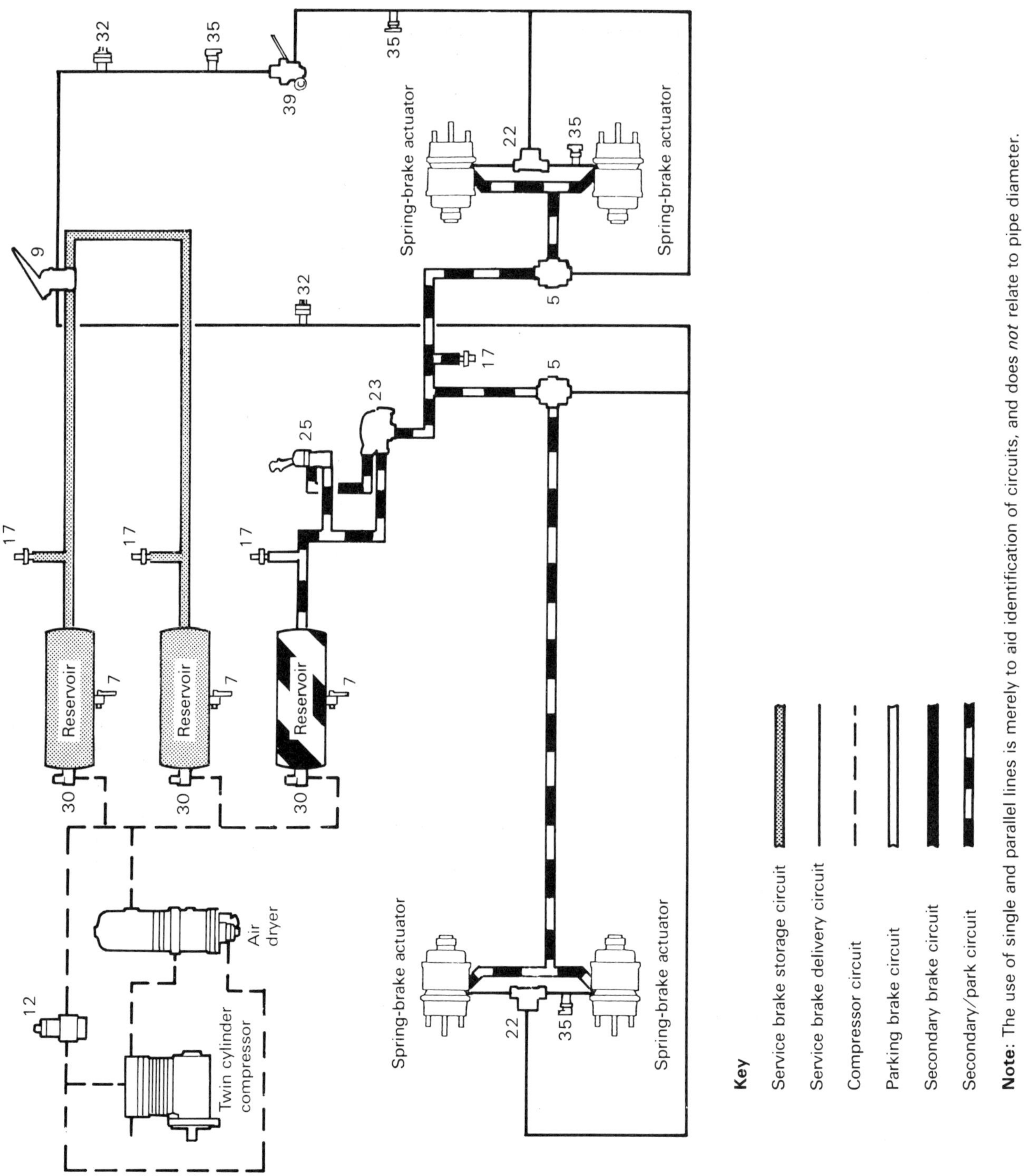

Figure 7.5 Truck – spring brakes

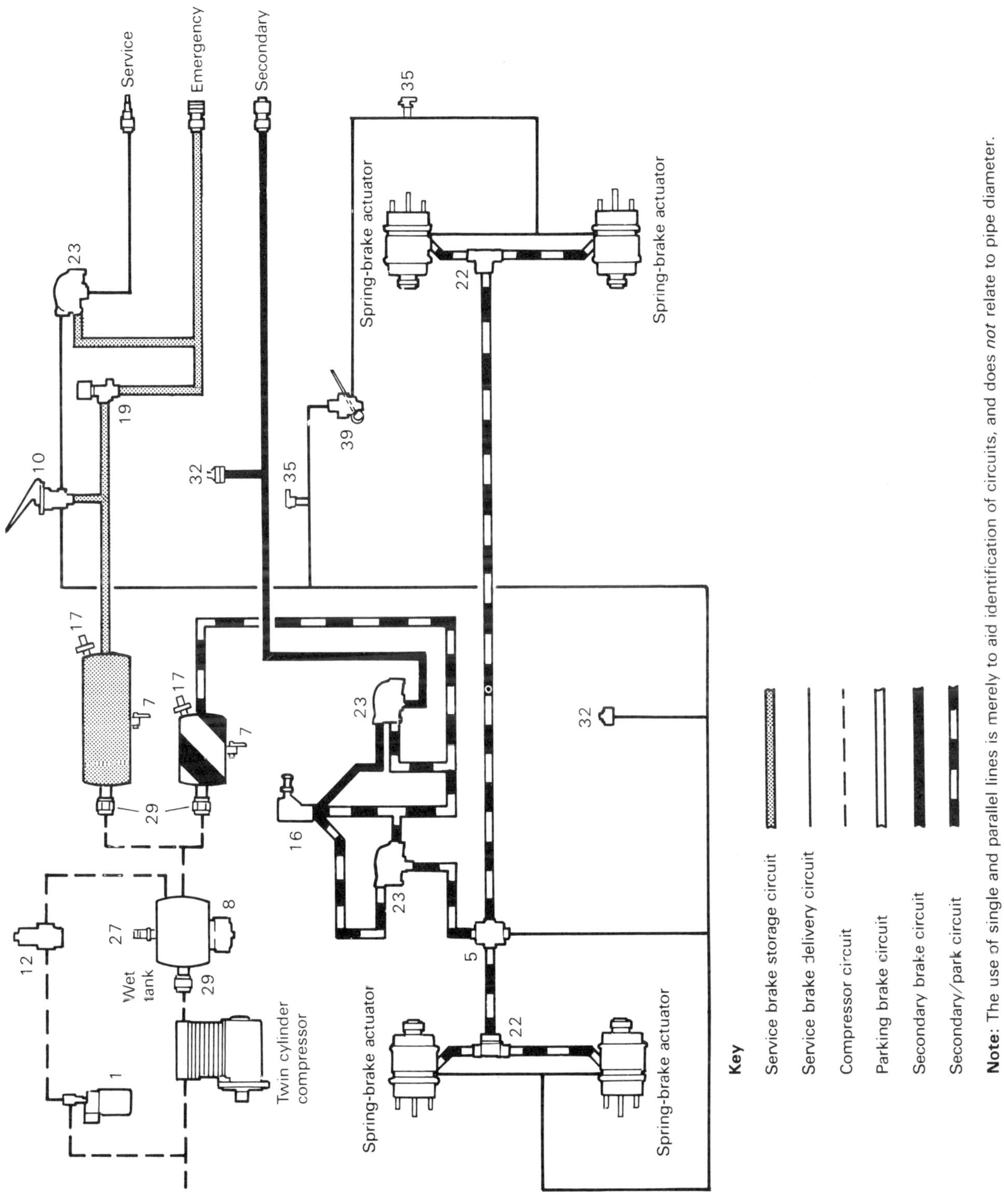

Figure 7.6 Tractor – spring brakes (1)

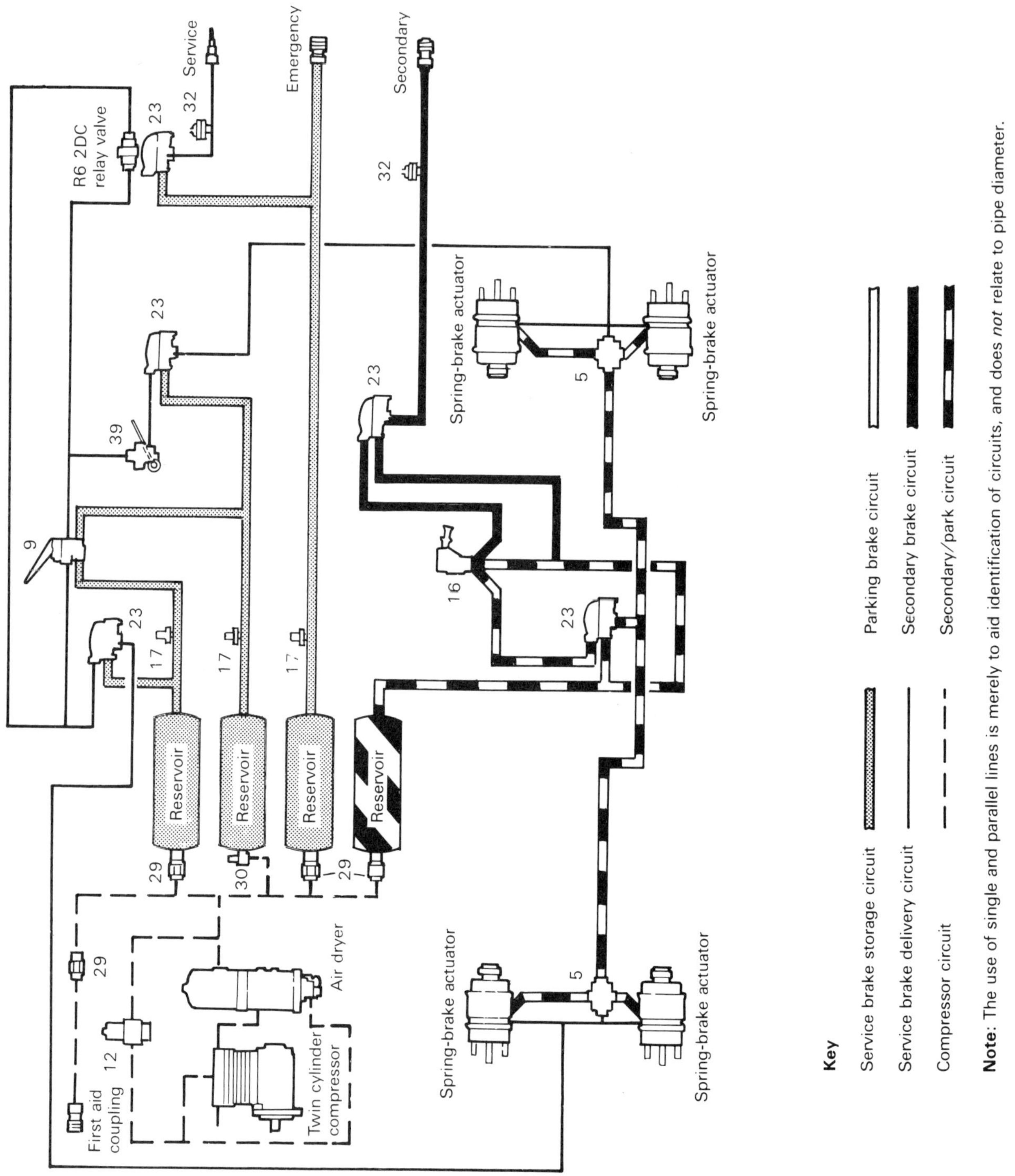

Figure 7.7 Tractor – spring brakes (2)

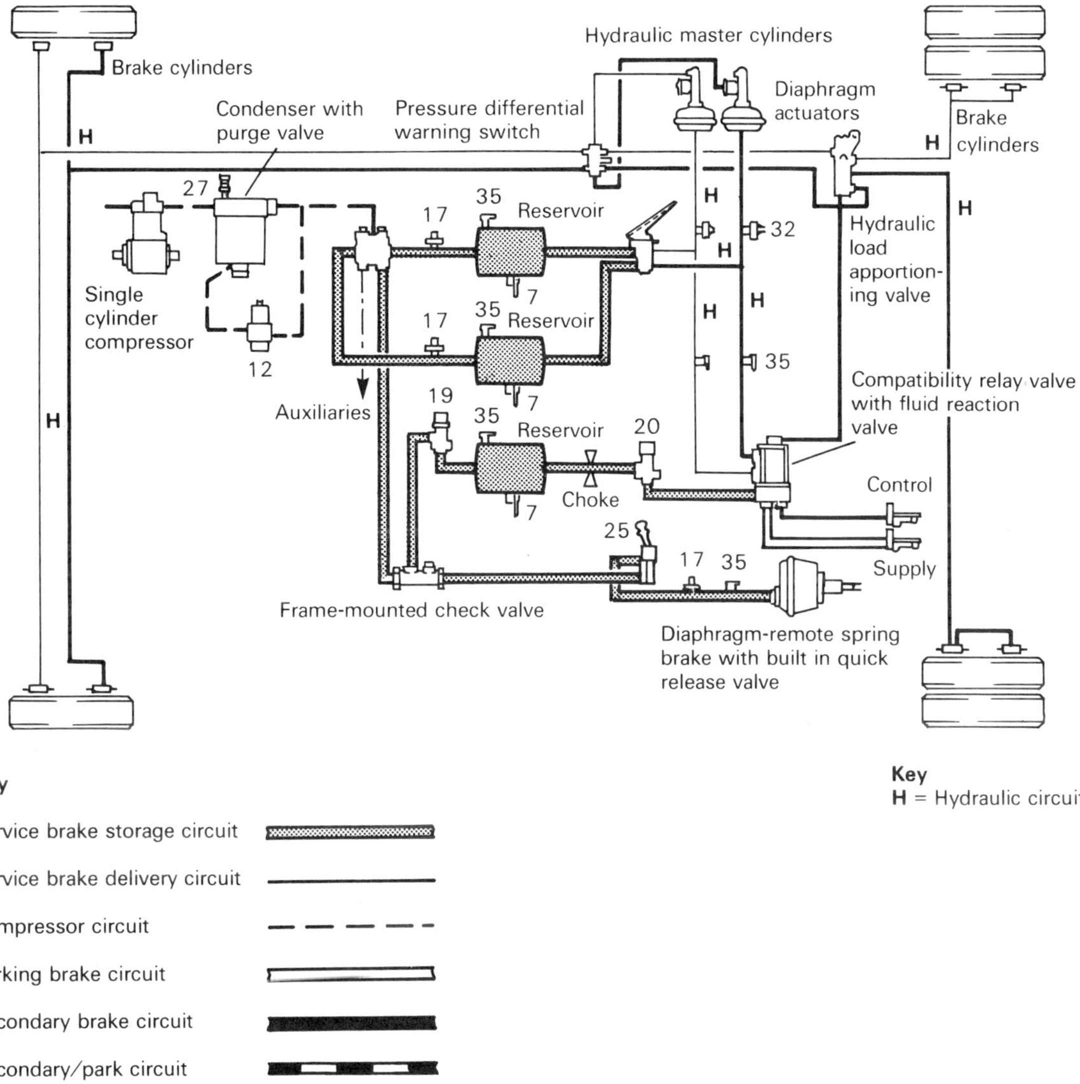

Note: The use of single and parallel lines is merely to aid identification of circuits, and does *not* relate to pipe diameter.

Figure 7.8 Air/hydraulic tractor

modern safeguards, including load-sensing valves. Operation of the foot valve causes compressed air to activate two independent air/hydraulic activators. One of these operates both front brakes and the rear nearside brake, whilst the other operates both front brakes and the rear offside brake (this is known as an *L-split*). If half of either the air or hydraulic systems fail, three out of four wheels will still be capable of braking. Beyond the air/hydraulic actuators the service brakes are hydraulically operated.

With this type of layout the footbrake system is classified as service and secondary because, in the event of a failure of half of the system the remainder is capable of a secondary performance. Parking is by a single spring-brake chamber, plus mechanical linkage.

Trailer braking is achieved by the use of a compatibility relay valve (with a fluid reaction valve). This also reacts to hydraulic signals from the load-sensing valve on the tractor to ensure braking compatibility.

Towing truck or tractor (1) (Figure 7.9)

This system has four reservoirs, two for the service system, one for trailer supply and one for the

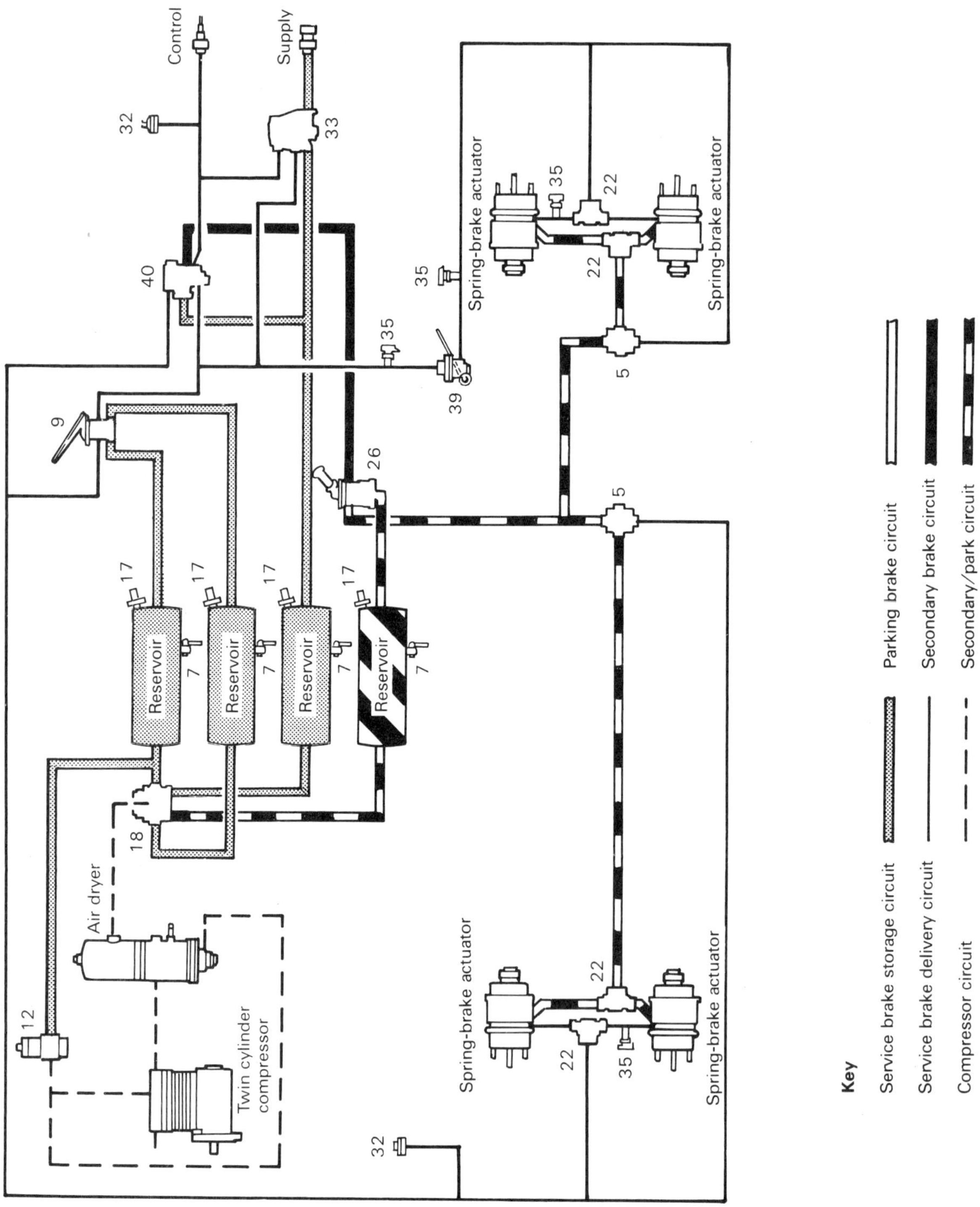

Note: The use of single and parallel lines is merely to aid identification of circuits, and does *not* relate to pipe diameter.

Figure 7.9 Towing truck or tractor (1)

secondary/park system. A multi-circuit protection valve is fitted to safeguard each individual reservoir and a dual foot valve is also fitted.

In this system, a hand valve with a straight gate is fitted. The secondary/park reservoir is necessary to comply with the regulations that require that the spring-brake circuit must include an energy reserve, which is not used to supply any other equipment, unless it can be supplied from at least two systems independent of each other. In addition, the emergency line is signalled from the parking supply, so that if the spring brakes come on automatically, the trailer brakes will also be applied.

The multi-relay valve (trailer-control valve) is necessary to conform with the requirement that if half the service system fails, the remaining section must be capable of applying the trailer brakes. The service, secondary and emergency lines all connect to this unit. The supply dump valve allows the trailer brakes to be applied automatically via the emergency (supply) line, if the service (control) line to the trailer is broken.

Two-line trailer (Figure 7.10)

This layout is suitable for semi-trailers or draw-bar trailers. A reservoir and a relay-emergency valve are installed as with a three-line trailer. However, if this system is fitted to a draw-bar trailer, it is necessary to provide a second variable-load valve to modulate the pressure applied to the brakes on the front axle, independently from those on the rear axle.

A release valve is fitted so that the trailer reservoirs can be drained, after the trailer has been disconnected from the unit, enabling it to be moved with the brakes released.

Towing truck or tractor (2) (Figure 7.11)

This three-axle tractor layout is fitted with twin wedge foundation brakes, which are operated by single-diaphragm (service) chambers and spring-operated piston chambers for the secondary and parking systems.

A dual foot valve controls the service brakes by a horizontally split system, i.e. one-half of the valve is connected to all the simple single-diaphragm chambers, whilst the other half controls the service chambers that are attached to the spring-brake chambers.

The secondary/park control is a straight gate hand valve, which causes the spring brakes on the tractor and the orthodox air brakes on the trailer to be applied simultaneously, when the lever is moved to the secondary position. This is achieved by exhausting air from the spring-brake chambers on the drawing unit, and also exhausting air from the inverse line to the trailer control valve, which causes the trailer brakes to be applied. The trailer control valve has two upright air connections and one inverse air connection.

This vehicle has two rear axles, but with the horizontal-split arrangement it is necessary to duplicate the load-sensing arrangement and, consequently, a slave variable-load valve is fitted in addition to a conventional variable-load valve.

A supply dump valve is used in this system. The requirement of this valve is that in the event of a failure of air pressure in the service (control) line to the trailer, within two seconds of the next *full* service brake application, the pressure will drop to 22 lbf/in^2 (150 kN/m^2, 1.5 bar), causing the relay emergency valve to apply the trailer brakes automatically.

Spring-brake trailer (Figure 7.12)

Service braking, on this type of trailer, operates the diaphragm (service) section of the spring-brake chambers via a relay valve.

The reservoir on this trailer lacks the conventional relay emergency valve, although the same emergency function is present through the use of an R61DC valve and an inversion valve. Operation of the secondary system exhausts pressure between the inversion valve and the signal port of the R61DC. This exhausts the pressure between the R61DC and the spring brakes, and cause these to apply. The inversion valve is supplied with compressed air by the red emergency (supply) line, and when disconnecting this hose, the single-check valve will retain air in the supply line to the relay valve, and automatic braking will occur as the supply from the inversion valve ceases.

When the trailer is parked and the red line is disconnected, it is possible to move the trailer after using the manoeuvring valve to release the trailer parking brakes.

American towing vehicle (Figure 7.13)

This diagram shows one system in common use on vehicles produced in the USA. The vehicle is fitted with spring brakes and is capable of drawing a trailer which is also fitted with spring brakes. An LQ4 front-axle valve regulates the service delivery pressure to the front brake chambers, when the foot valve is lightly applied.

The SR1 spring-brake valve is signalled by the front-brake delivery pressure and the rear-axle supply pressure. As a result, it supplies a hold-off pressure to

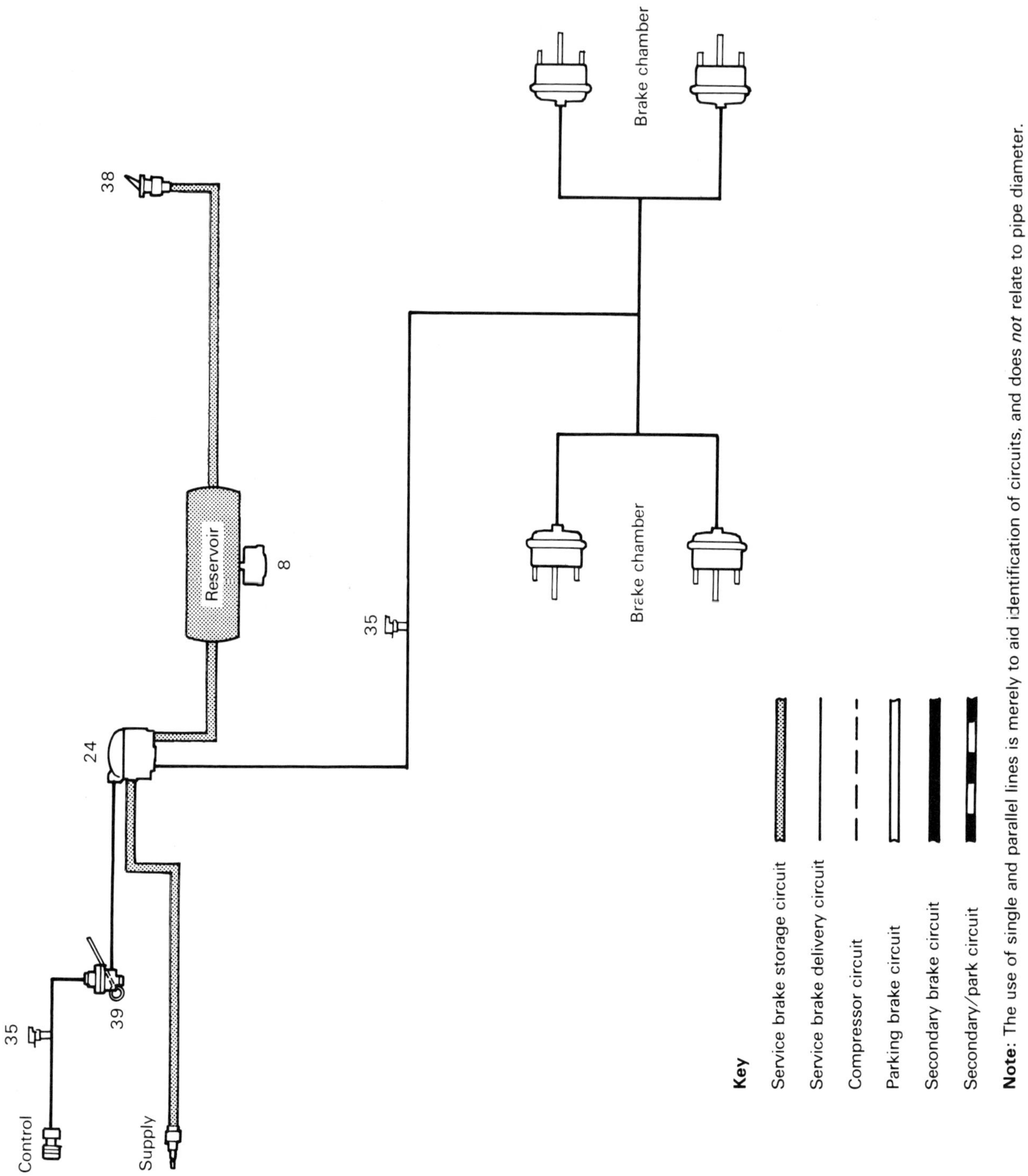

Figure 7.10 Two-line trailer

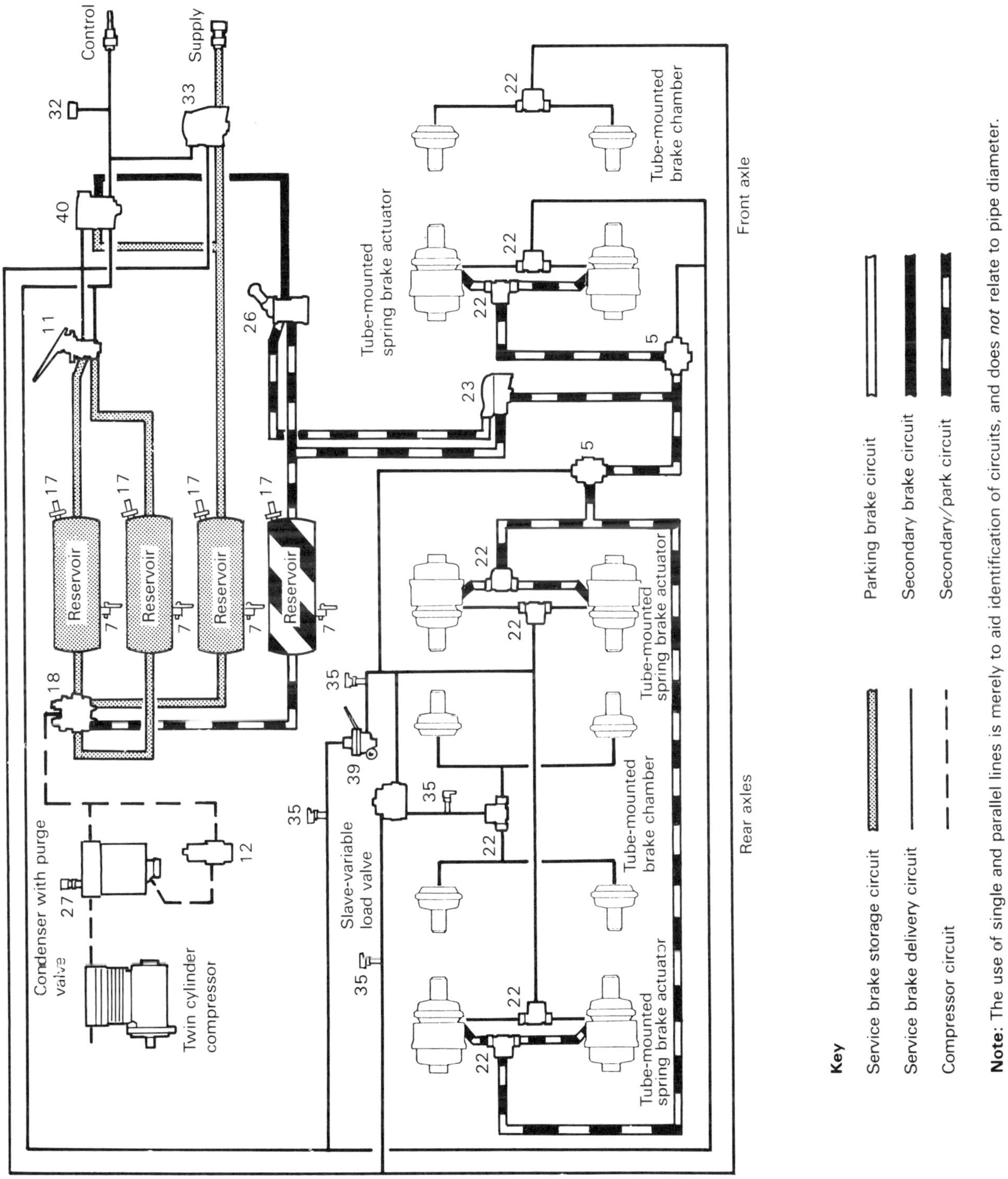

Figure 7.11 Towing truck or tractor (2)

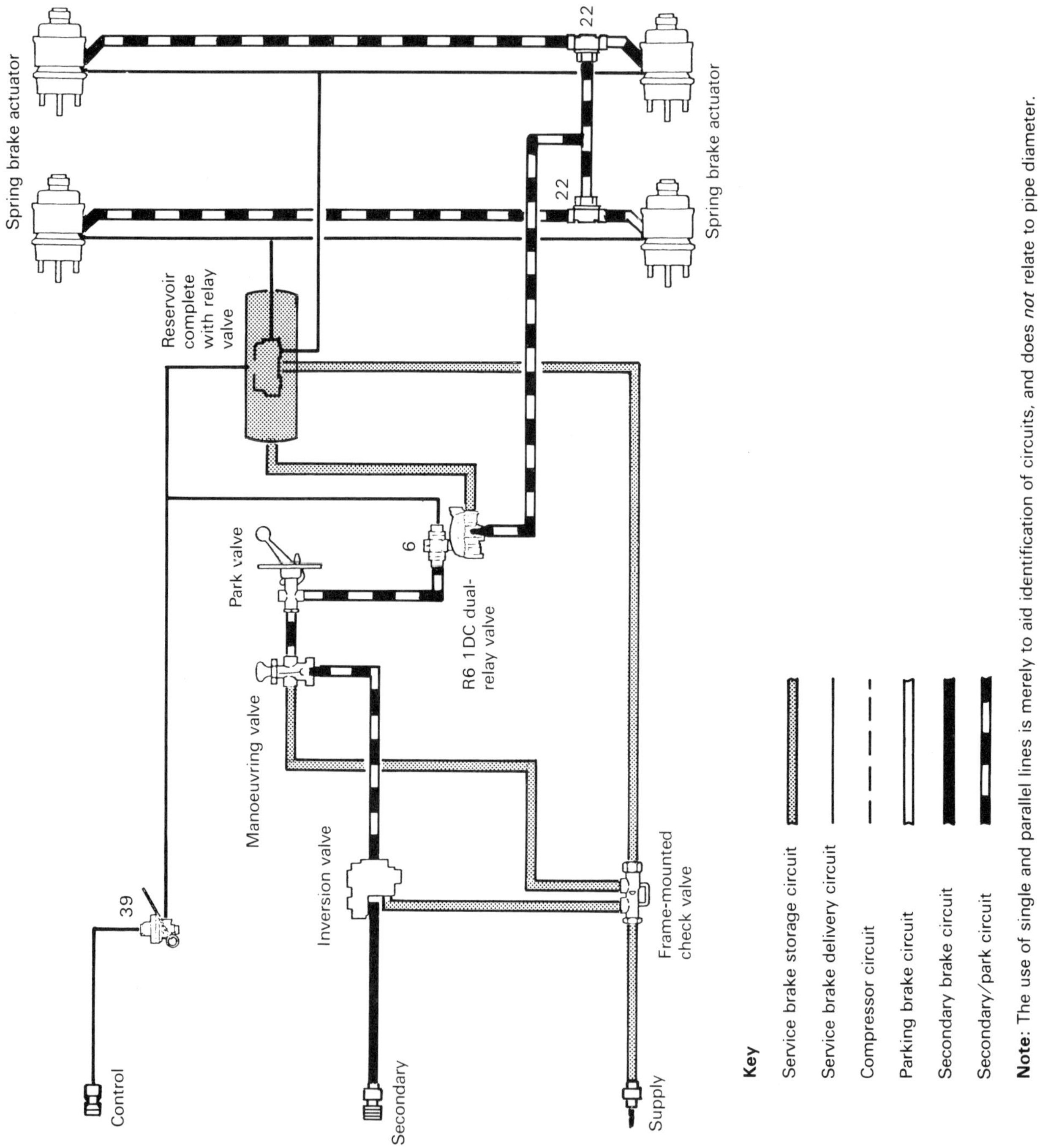

Figure 7.12 Spring-brake trailer

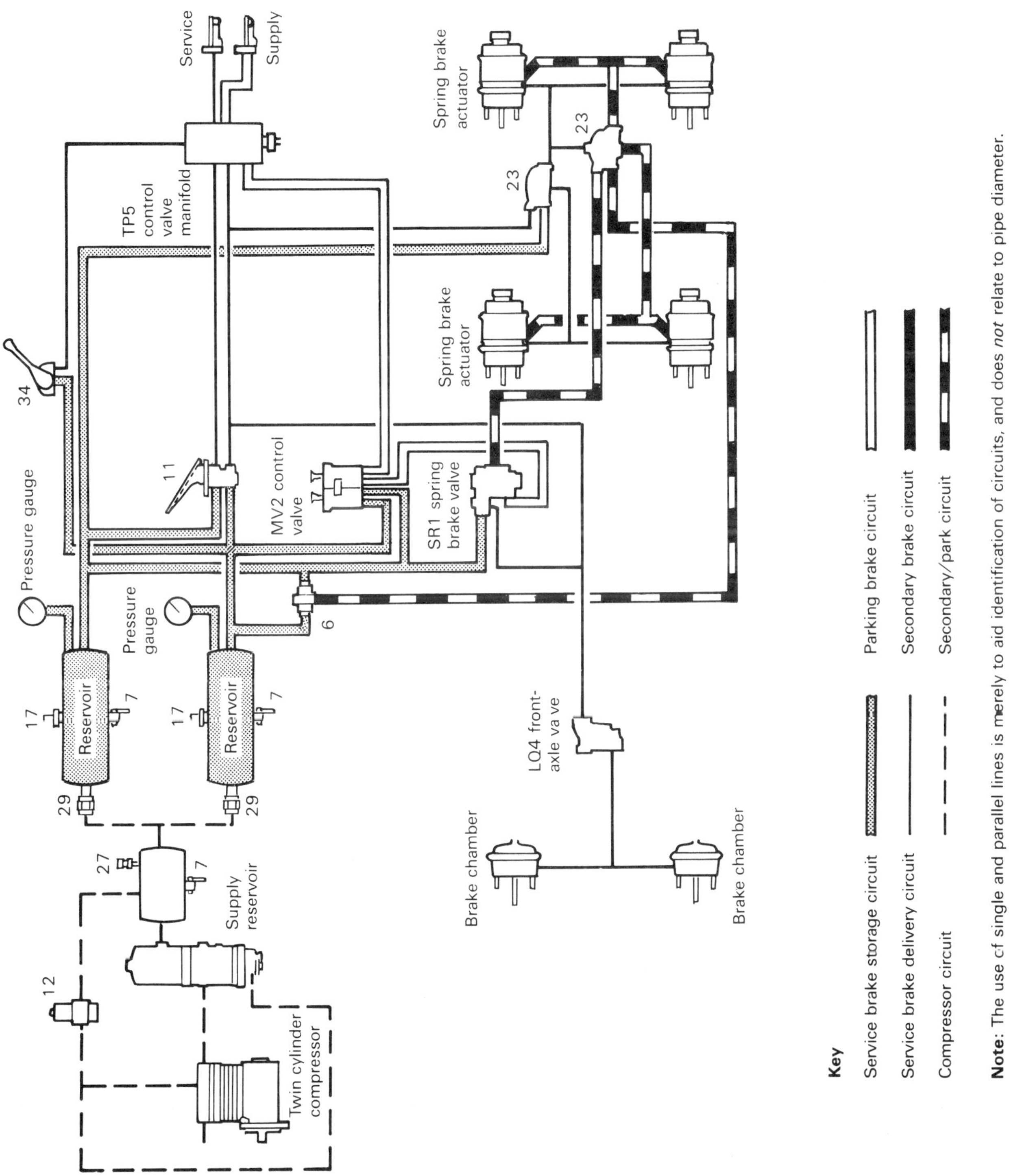

Figure 7.13 American towing vehicle

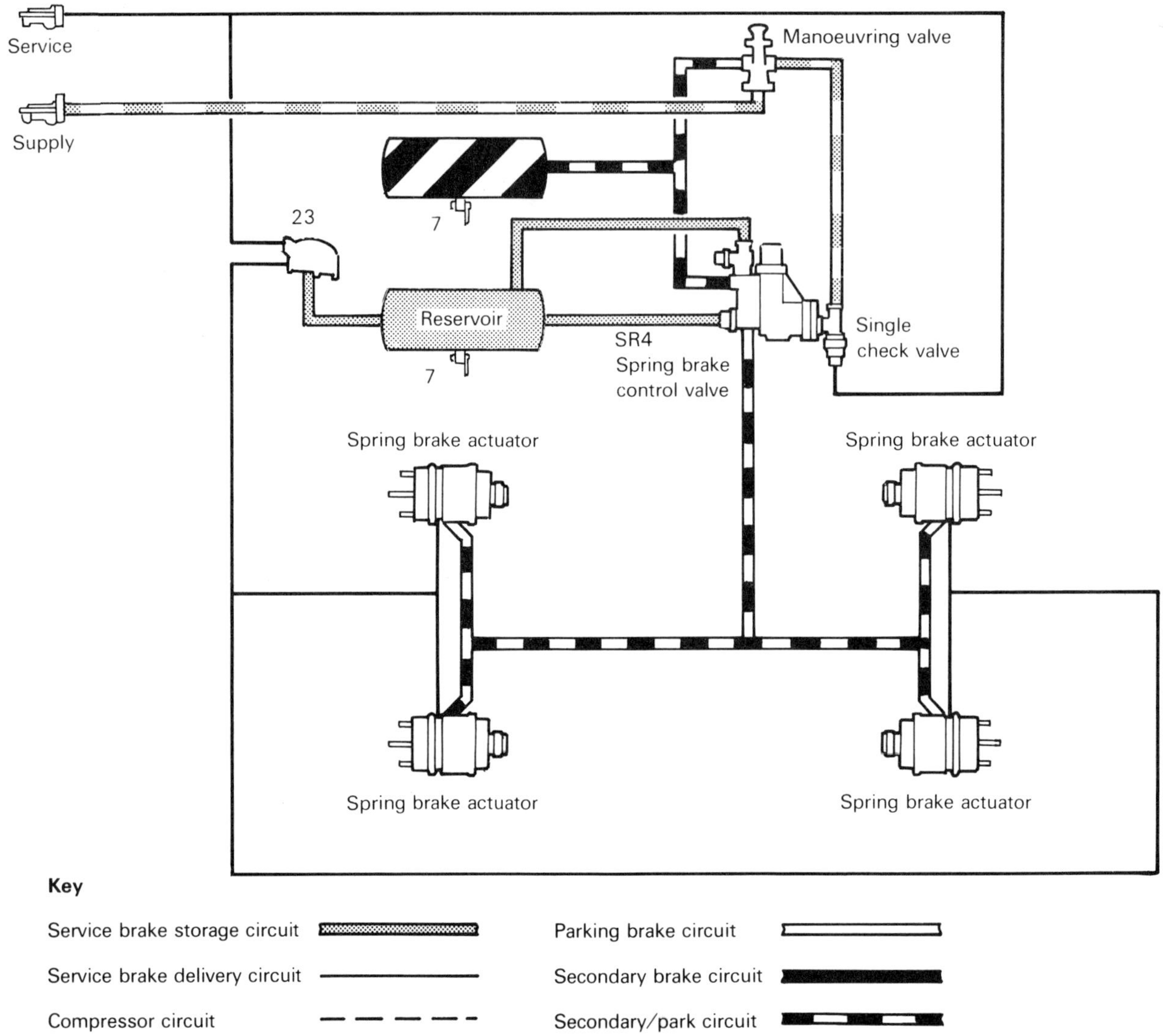

Note: The use of single and parallel lines is merely to aid identification of circuits, and does *not* relate to pipe diameter.

Figure 7.14 American trailer

the spring-brake chambers. In the event of a serious pressure drop in the reservoirs that supply the rear-axle brake chambers, this valve will reduce the hold-off pressure to the spring brakes, in proportion to the pressure that is applied to the service-brake chambers on the front axle.

A driver control, the MV2 dashboard/control panel mounted valve, consists of one red and one yellow pull/push knob. The red knob affects the trailer supply line and the yellow knob affects the trailer spring brakes.

An R8 relay valve, which incorporates an anti-compound device is fitted in this system.

American trailer (Figure 7.14)

The system shown is very similar to an EEC two-line trailer. It incorporates a manoeuvring valve to aid the recovery of the trailer in the event of serious defect, or to allow the normal manoeuvring of a detached trailer in the course of transport operations.

The SR4 spring-brake control valve has a somewhat similar role as the relay emergency valve. A single check valve is used as an anti-compound device.

Figures 7.15 to 7.18

Figures 7.15 to 7.18 show examples of piping diagrams for various vehicles in current use.

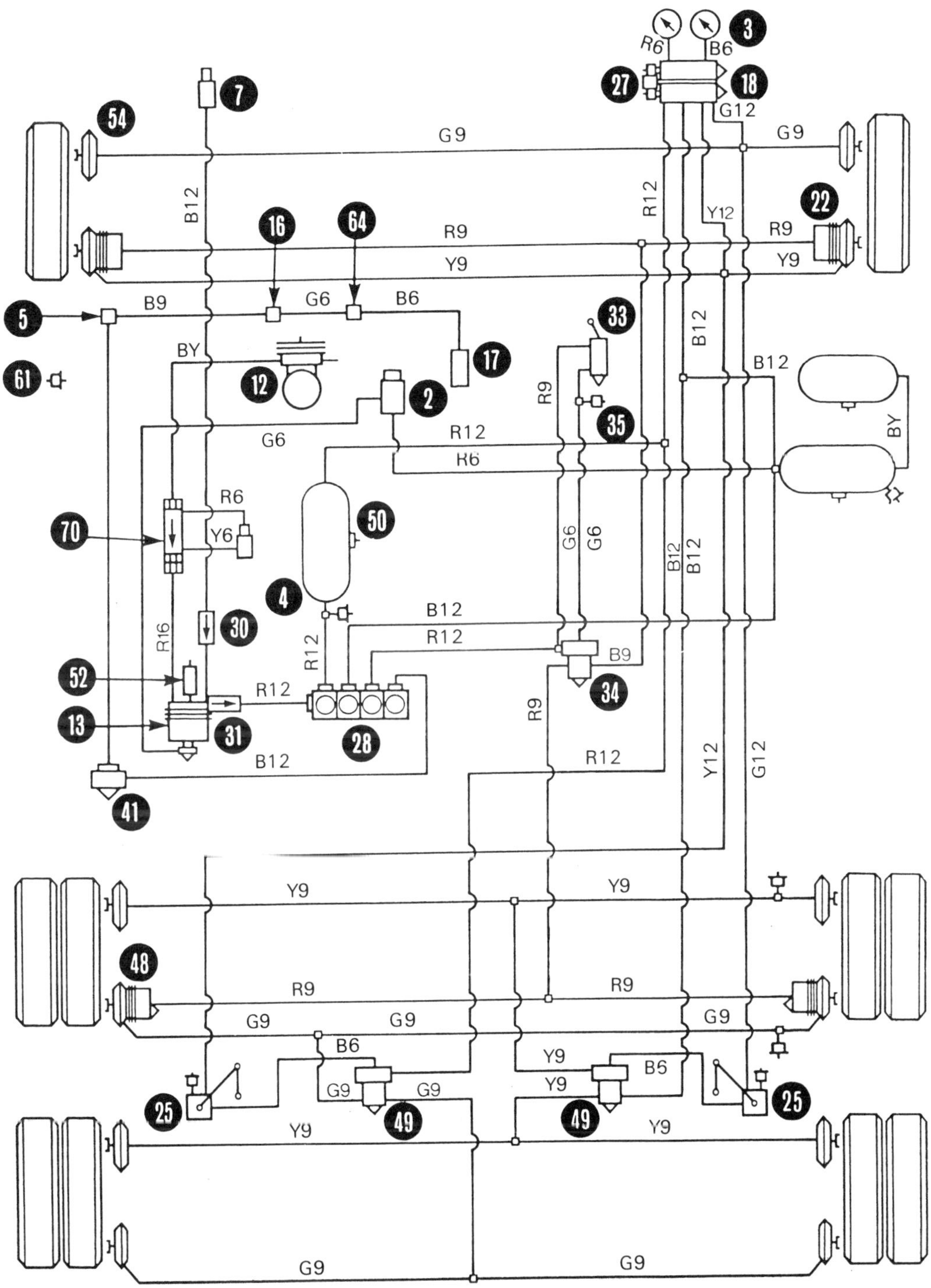

B = Blue = Parking brake circuit
R = Red = Air supply circuit
Y = Yellow = Service brake circuit
G = Green = Trailer brake circuit
BY = Bundry tube

Figure 7.15 Iveco Ford truck

Key

2 = Air limit valve
3 = Air pressure gauges
4 = Alcohol injector
5 = Auxiliaries seven-way junction
7 = Blow up connector
12 = Compressor
13 = Condenser and drain valve
16 = Exhaust brake control valve
17 = Exhaust butterfly valve
18 = Foot control valve
22 = Front spring actuator
25 = Load-sensing valve
27 = Low air pressure and stop light switches
28 = Multi-circuit protection valve
30 = Non-return valve
31 = One-way valve
33 = Parking brake control valve
34 = Parking brake relay valve
35 = Parking brake warning light switch
41 = Pressure reduction valve
48 = Rear spring actuator
49 = Relay valve
50 = Reservoirs
52 = Safety valve
54 = Service actuator
61 = Test points (where fitted)
64 = Throttle shut-off valve
70 = Check valve

Key

1 = Brake footvalve
2 = Air gauges
3 = Low air and stop lamp switches
4 = Air reservoir - front service brake
5 = Spring brake emergency release coupling
6 = Relay valve - front service brake
7 = Air compressor
8 = Spring brake hand control valve
9 = Spring brake chamber
10 = Differential lock switch
12 = Emergency line coupling
13 = Service brake line coupling
14 = Auxiliary supply manifold
15 = Governor
16 = Differential protection valve
17 = Manifold block
18 = Check valve
20 = Relay valve - spring brakes
21 = Air dryer
22 = Multi-circuit protection valve
23 = Hose connection - differential lock
24 = Quick-release valve
25 = Spring brake chamber
26 = Variable-load valve
27 = Relay valve - rear service brake
28 = Air reservoir - trailer and parking
29 = Air reservoir - rear service brake

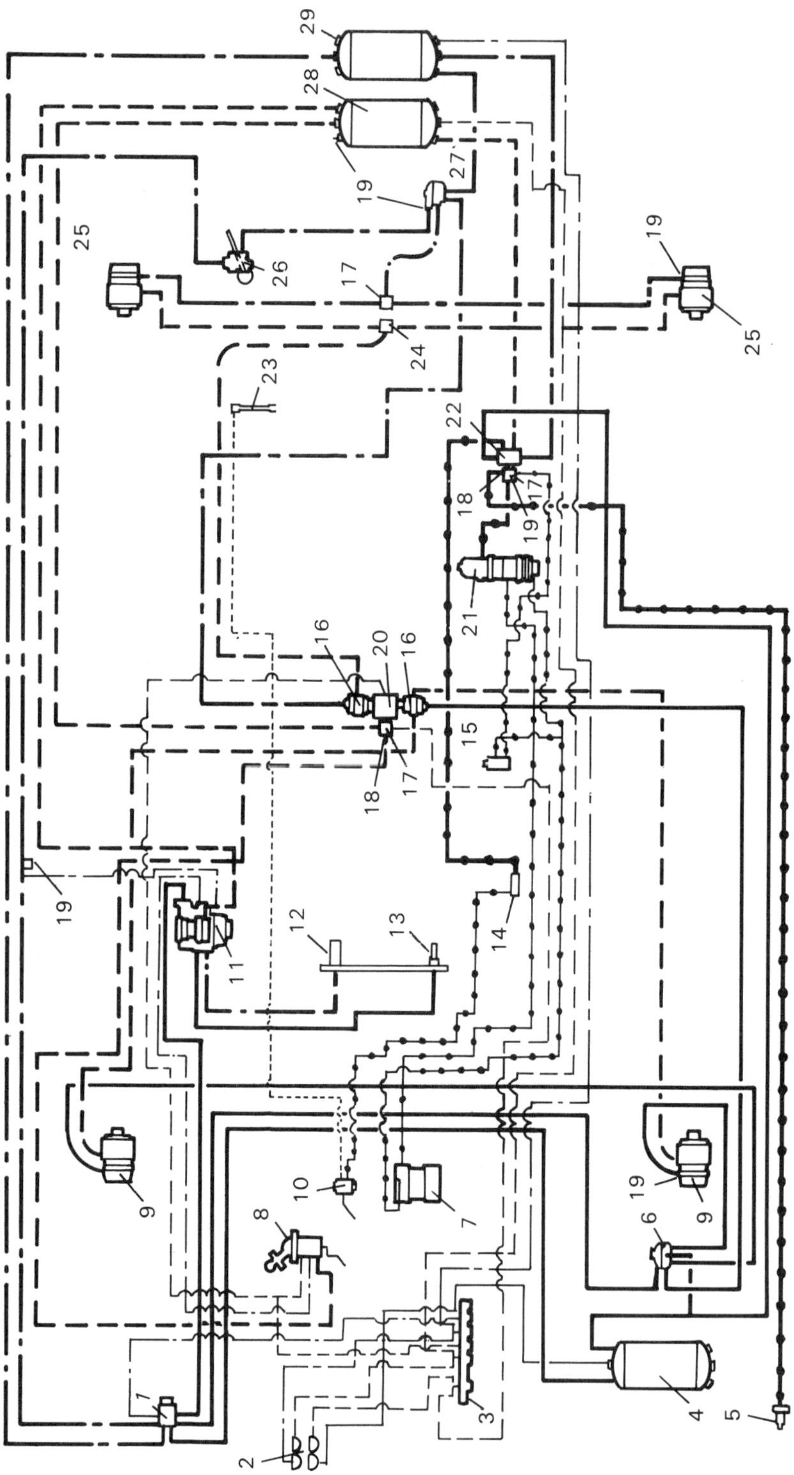

Key

Pump-up lines, auxiliary lines and governor, compressor and air dryer signal lines

Differential lock feed line

Front axle and trailer service lines

Rear axle service and trailer emergency lines

Spring brake lines

Figure 7.16 ERF tractor

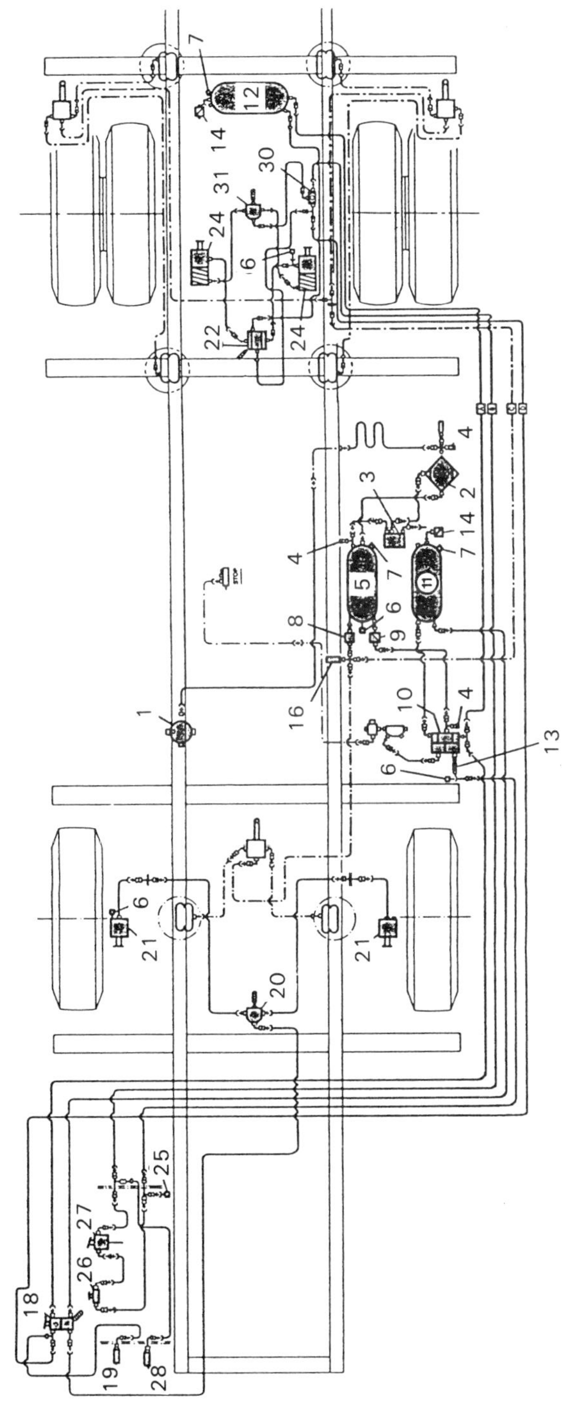

Key

1 = Compressor
2 = Air dryer
3 = Regulator
4 = Safety valve
5 = Wet tank
6 = Test point
7 = Drain valve
8 = Overflow valve for air suspension system
9 = Pressure limiting valve
10 = Multi-circuit protection valve
11 = Reservoir
12 = Reservoir
13 = Non-return valve
14 = Sensor for air gauge
16 = Low-pressure indicator for air suspension system's warning lamp
18 = Footbrake valve
19 = Brake lamp switch
20 = Quick-release valve, front-axle brake
21 = Brake cylinder, front axle brake
22 = Relay valve
24 = Spring brake cylinder, drive shaft brake
25 = Inflation point
26 = Blocking valve
27 = Parking brake valve
28 = Low pressure indicator for brakes
30 = Two-way valve
31 = Quick-release valve, parking brake

Figure 7.17 Volvo Leyland coach

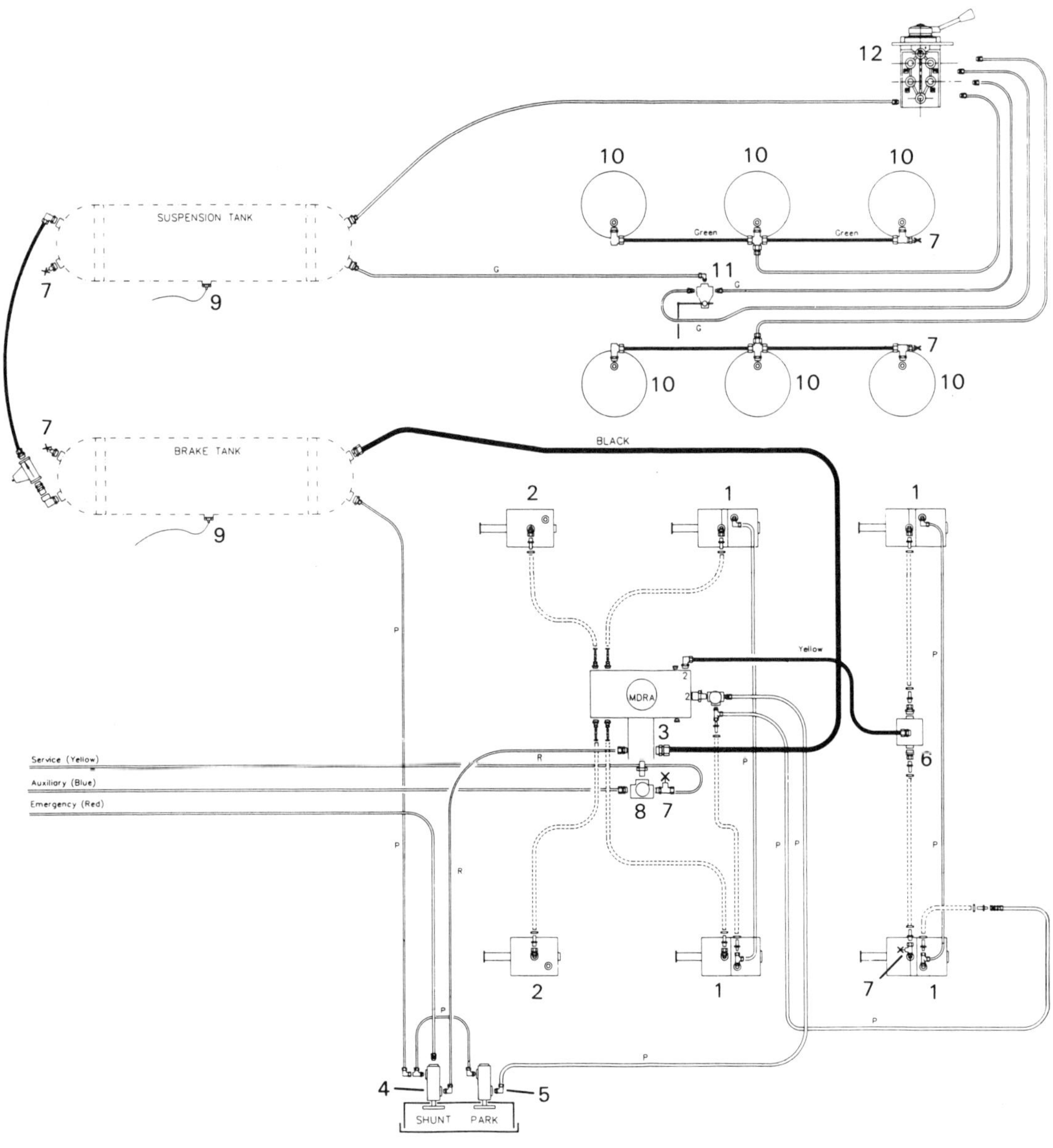

Figure 7.18 Tri-axle semi-trailer (YK)

Key

1 = Spring brake chamber
2 = Brake chamber
3 = ABS
4 = Shunt valve
5 = Park valve
6 = Quick-release valve
7 = Test point
8 = Double-check valve
9 = Drain valve
10 = Suspension bellows
11 = Levelling valve
12 = Raise/lower valve

Colour code for tri-axle semi-trailer (YK) (Fig. 7.18)

Blue = Auxiliary
G = Green = Air suspension
P = Purple = Spring-brake release circuit
R = Red = Emergency
Yellow = Service
Black = Supply to anti-lock

8 Safety checks

Regular maintenance will give early warning of a large number of symptoms of failure that can affect safety, convenience and cost.

Basic safety precautions

Basic safety precautions, when working on an air brake system, should include:

1 Stop the engine before working under a vehicle.
2 Always chock vehicle wheels.
3 Keep hands away from actuator push-rods and slack adjusters as these may apply as system pressure drops.
4 Never connect or disconnect a hose line containing air pressure, it may whip as air escapes.
5 Never remove a component or pipe plug unless you are certain all system pressure has been depleted.
6 Never exceed recommended air pressure.
7 Wear safety glasses when working with air pressure.
8 Never look into air jets or direct them at anyone.
9 Never attempt to dismantle a component until you have read and understood recommended procedures. Some units contain powerful springs and injury can result if not properly dismantled. Use only correct tools and observe all precautions relative to the use of these tools.

Drain cocks or plugs on reservoirs should be opened regularly to prevent moisture reaching other parts of the braking system and causing deterioration. In cold weather, such moisture can freeze, causing damage or failure.

To check over the whole system, first chock the road wheels, then drain all reservoirs. Check the oil level in the compressor, unless it is fed by the engine. Make sure that hand and foot valves are in the 'off' position. Now check that all drain cocks are closed. Turn on the ignition switch. The low-pressure warning buzzer should sound immediately. Start the engine and then check the gauges to ensure that the pressure is rising. If individual check valves are fitted to both service and secondary reservoirs, the pressure should rise evenly. However, if the secondary reservoir is fed via the pressure-regulator valve or charging valve, the service reservoir will rise to the valve setting before the secondary gauge needle leaves zero.

Note: Some current vehicles are, and all future vehicles will be, fitted with some form of system-protection valves (also called circuit-protection valves). These valves cause the reservoirs to charge up one at a time, and a particular note should be made of opening differentials between the first and last reservoirs during the charging up of the system. These can be checked against the manufacturer's data if in doubt. (Some of these valves have a feedback capability to equalise pressure.)

The buzzer should continue sounding until all the pressure gauges show between 60–70 lbf/in^2 (414–483 kN/m^2, 4.0–4.8 bar). With the engine running, just below maximum governed speed, the air pressure in the system should rise from zero to the low-pressure setting (i.e. stop the buzzer) in three minutes. (At present, Department of Transport test centres in this country require this performance from both rigid vehicles and combinations.)

Continue to build up the pressure until the unloader valve starts discharging to atmosphere, then switch off the engine. The reading on all the gauges should remain steady, but a tolerance of 1 lbf/in^2 (7 kN/m^2, 0.07 bar) per minute (maximum) loss is allowed. However, there should be no leaks from pipework or fittings. Now, fully apply the footbrake and keep the foot hard down. The loss in the pressure should not exceed 2 lbf/in^2 (14 kN/m^2, 0.1 bar) per minute. Release the footbrake and recheck the reservoir pressures. Repeat these checks on the secondary system by using the hand valve. Check that the stop lights operate for both service and secondary circuits. Soapy water is the best leak finder.

Spring brakes

Check the footbrake as above. Then ensure that the reservoirs are fully charged and stop the engine. Put the handlever in the park position (the gauge will not show a drop in pressure) then bring the lever across through the gate to the 'off' position. A drop in pressure should be noted and the brakes should be free.

Trailers

Semi- or draw-bar trailers are fitted with a breakaway device (usually a relay emergency valve). This can be checked by disconnecting the emergency line Susie hose, which should cause the trailer brakes to fully apply. Trailer brakes should remain on until 60–70 lbf/in^2 (414–483 kN/m^2, 4.0–4.8 bar) is reaching the relay emergency valve. A pressure-protection valve on a tractor should ensure that the tractor emergency line will *not* feed the trailer brakes, only the tractor, if the pressure drops to 45–50 lbf/in^2 (310–354 kN/m^2, 3.0–3.5 bar). Check that no leaks occur at the relay emergency valve when the hand valve is used.

Tracing faults

Fault finding on compressed air equipment can be compared with electrical fault finding as it involves more thought than action.

Systematic checking is essential, unless the fault is obvious. The four parts of the system, compression, storage, control and actuation, should be carefully considered, and a decision made as to which part of the system to check. For example, if there is a drop in gauge pressure when the handvalve is applied, which does not occur at other times, it would be pointless to check the compression section of the system, or indeed storage or actuation. Depending upon the type of defect eventually found, or simply to ensure that the whole of the braking system is kept in good order, it may be prudent to check the rest of the systems, *after* tracing and curing the original defect.

What follows is one method of locating faults.

Test 1 – compression

Empty all the reservoirs. Run the engine at just below governed speed for 3 minutes and observe the cab gauges. If the gauges reach a safe pressure (i.e. the low-pressure warning device ceases to operate), move on to check the storage system (if this is in doubt). If only one or two of the three gauges (an average number fitted) reach a low reading, whereas one or more reach a normal reading, check the storage system. If all three gauges take an excessive time to build up, the defect could be in either the compression or the storage system.

Assuming that the gauges have not reached the required pressure, a check for leakages using soapy water should be made, with the engine still running, at the compressor, unloader or governor (especially at exhaust ports), condenser or wet tank, automatic drain valve (particularly at the drain port), low-pressure switch (if fitted in this part of the system), and all pipes, unions or connections. Check that the intake filter to the compressor is not blocked. If no pressure is registered on any of the gauges, check that the mechanical drive for the compressor is connected and/or tight. If these checks fail to find a defect, use a test gauge to check the output from the compressor and the functioning of the unloader or governor. A defective compressor should be dismantled and overhauled, whereas the unloader may only need adjusting rather than overhauling. (A cab air pressure gauge may be defective.)

Test 2 – storage

Build up the pressure to unloader or governor cut-out pressure, if possible, and then switch off the engine. With the reservoir fully charged, the engine stopped and no control valves in use, the maximum leakage that can be tolerated is 20 lbf/in^2 (138 kN/m^2, 1.4 bar) in 20 minutes, for a rigid vehicle, or 30 lbf/in^2 (207 kN/m^2, 2 bar) in 20 minutes, for a combination, (i.e. a drawing vehicle and trailer). If the leakage is in excess of these limits, check the following components for leakage. As the engine is not running, this may be audible, or it may be traced by using soapy water. These units are: reservoirs, their drain cocks and non-return valves (both on the unit and the trailer), foot valve, hand valve, relay valves, relay emergency valve, emergency Susie hose, all pipelines and connections with the circuit.

Note: Although this section is classified as storage, reservoir pressure is always present up to the inlet port of control valves, relay valves and relay emergency valves, as well as in the emergency line/Susie hose.

Loss could also occur at the check valve within the unloader valve or from the low-pressure switch, although these should have been found if there had been a need to check the compression circuit.

There could well be other units, for example a pressure-protection valve or pressure-regulator valve, which are fitted to the particular vehicle and also have reservoir pressure present at their inlet port. Therefore, these should also receive attention. The manufacturer's workshop manual should give relevant details, if the person doing the job is not familiar with the particular vehicle. Leakage from the exhaust port of the valves, when the control valves are not in use, indicates that the inlet valve is defective, as the inlet should be closed and the exhaust open. (Always remember that a cab air pressure gauge may be defective.)

Test 3 – control

Both the foot and the hand control valve should be applied (independently), and then held applied, before being released. The engine should not be running for these tests. Test all components in these circuits for leaks.

Foot valve: Check the stop-light switch for leaks. Check the variable-load valve, or the load-sensing valve, especially at the exhaust port. The brake chamber should be checked. This check can include putting masking tape over the drain holes in the non-pressure plate, so that any escaping air is concentrated and easier to detect. The service line, including the Susie hose, should be checked as should all pipe connections. The relay emergency valve should be checked, especially the exhaust port, which is on top of the RE4 and at the base of the RE6. Relay valves should be checked in a similar way.

Hand valve: Relay valve, quick-release valves on both the unit and the trailer, stop-light switch, brake chambers (to check the secondary diaphragm), all piping, the secondary Susie hose, all connections, etc., should be checked for leaks, with the control valve applied and again with it released.

Leakage: With control valves (for example, foot or hand valves), a leak at the exhaust valve, when that control is not applied, indicates an inlet valve failure. A leak at the exhaust valve, when the control is applied, indicates an exhaust valve failure.

A suggested tolerance in braking systems for rigid vehicles is as follows:

> with systems fully charged and the engine stopped, and where brakes are *not* applied, approximately 1 lbf/in^2 (7 kN/m^2, 0.07 bar) per minute; and *with* brakes applied, approximately 2 lbf/in^2 (14 kN/m^2, 0.1 bar) per minute.

When dealing with a combination of vehicles, for example an articulated vehicle, a slight increase in the leakage rates can be allowed.

Test 4 – actuation

This can be seen by observation on the push-rods of brake chambers and by removing drums. Lack of adjustment or lubrication, weak or broken return springs and worn linings, form the basis of faults in this section.

The following are a selection of specific faults that drivers report to workshop staff and the corrective action that should be taken in each case.

Brakes grabbing: Ensure that pneumatic equipment is functioning correctly. Check the brake drums for ovality, grease or an excessive amount of dust on the linings, and for weak return springs. A lack of lubrication on linkage can also cause this defect.

Brakes slow to release: Check that the control valves are fully returning, there is a possibility of weak springs or lack of lubrication. Verify that the exhaust ports of the relevant valves are not obstructed. Observe the push-rod action as the defect could be due to weak return springs, either in the brake chamber or the brake shoes.

Uneven braking: If the vehicle is pulling to one side, check that the brakes are operating correctly on the other side of the vehicle. Likely possibilities are an air leak or grease on linings that cause that side of the vehicle to be ineffectively braked. Alternatively, there may be a defect in the pneumatic system to one side of the vehicle.

Brakes do not release: Check that the control valves are fully returning, as weak springs or lack of lubrication within this unit could cause such a defect. Watch the brake chamber push-rod action as lack of movement could be due to lack of lubrication or a broken return spring. The brake shoe return springs could be weak or broken. The exhaust ports of the various valves could be blocked. Pipelines could be restricted.

Brakes not adequate: Check the push-rod travel. The brakes may need adjusting or relining, or the linkage may need lubrication. Check for air leaks in the system with each control valve applied in turn. Using a test gauge, check the output from the control valves. Providing this reading is satisfactory, check the other valves in the system until the defect is traced.

Brakes do not work: Check which system, service or secondary, is at fault and also whether this failure occurs in a laden or unladen condition, or in both. Check the push-rod travel. If the rods are travelling the normal amount the defect will probably be with the foundation brakes. If the rods are not moving or moving to a lesser extent than normal, consider failures in the pneumatic system. Check the cab gauge readings and if these are low apply Test 2 on the storage system. The result of this could even cause Test 1 on the compression system to be applied. However, assuming that Tests 1 and 2 provide good results, move on to Test 3 on the control system, correcting any defects that are found.

It is not suggested that the above tests are the only or best methods available but that some method is better than none at all.

9 Testing, test equipment and maintenance

Most large companies that maintain air brake vehicles have developed sound maintenance schemes. However, not all of these companies have avoided the prohibition notice from the Department of Transport, still often referred to as GV9. This notice is currently called PG9, as it now covers both heavy goods vehicles (HGVs) and public service vehicles (PSVs). There is no magical solution to avoiding a prohibition notice, but as braking systems become more complex, we could be approaching a time, in theory at least, when it would be useful to keep one person, or group of people in the larger workshop, that specialises in pneumatic brake equipment.

Many companies do not recondition valves for braking systems. There are three important reasons for this attitude:

1 By just fitting an exchange unit there is a reduction in the 'vehicle off road' time.
2 The overhaul of certain complicated valves is a lengthy procedure and, therefore, the labour cost is high.
3 There may be a nagging doubt whether the job will be successful.

The first two reasons are obviously best assessed by the company concerned, but the third can be tackled by selecting *one* person to overhaul at least some valves. It is an open secret that the delivery time for certain valves is a little slow and that the waiting time is a problem for both haulier and manufacturer. One approach to this problem is that the next time a valve is replaced in a workshop, a new valve is fitted and the vehicle returned to service, but that the old valve is not used for exchange. When there is a slack period, the old valve can be reconditioned and then put into stock. The cost of a reconditioning kit is considerably less than that of an exchange valve. A typical example is the spring-brake chamber, but of course special tools and *care* are needed in this particular case. Even so, the initial outlay for the necessary press is soon recovered with a reasonably-sized fleet.

The type of valve to start with would be a quick-release valve. This simply requires the replacement of an O-ring and a diaphragm, which is both a quick, simple job and a morale booster. Manufacturers produce excellent literature, covering overhaul procedures, which is readily available.

The following pages list the types of schedules that manufacturers suggest for their equipment. It is not possible to cover all valves or indeed the recommendations of two manufacturers of the same type of valve. Some of the schedules are laid down by one company and some by others. The intention is to show some of the pitfalls of certain valves and the simplicity of other valves. This list is not intended to replace any schedules, merely to give a rough guide to be backed up by service literature.

Note: Where the annual mileage figure or 50 000 miles is shown, maintenance should be carried out at whichever figure is reached first.

Testing pneumatic equipment

With the advent of fail-safe systems, the problem of how to recover vehicles fitted with such equipment, if the engine and, consequently, the compressor cannot be run to charge the system to release the brakes, has arisen. With some types the brakes can be wound off, but with others, compressed air is needed even to move the vehicle. Test points are increasingly being fitted by manufacturers and some users have modified older vehicles. This, in some cases, allows the reservoirs to be charged from another supply, even from a commercial vehicle's tyre with its high pressure and large volume. However, a dangerous situation can develop if the vehicle is then driven back to base without correcting the fault. In the workshop the advent of these test points simplifies the testing of the air brake equipment. A common use for these is to check the input and output pressure from a variable-load or load-sensing valve, which is designed to reduce the braking available at the rear wheels of an unladen vehicle.

Scrutineer (HE)

Within the workshop, it is often both inconvenient and annoying to repeatedly charge the reservoirs by running the engine, merely to check the braking system. As a result, a number of devices have been

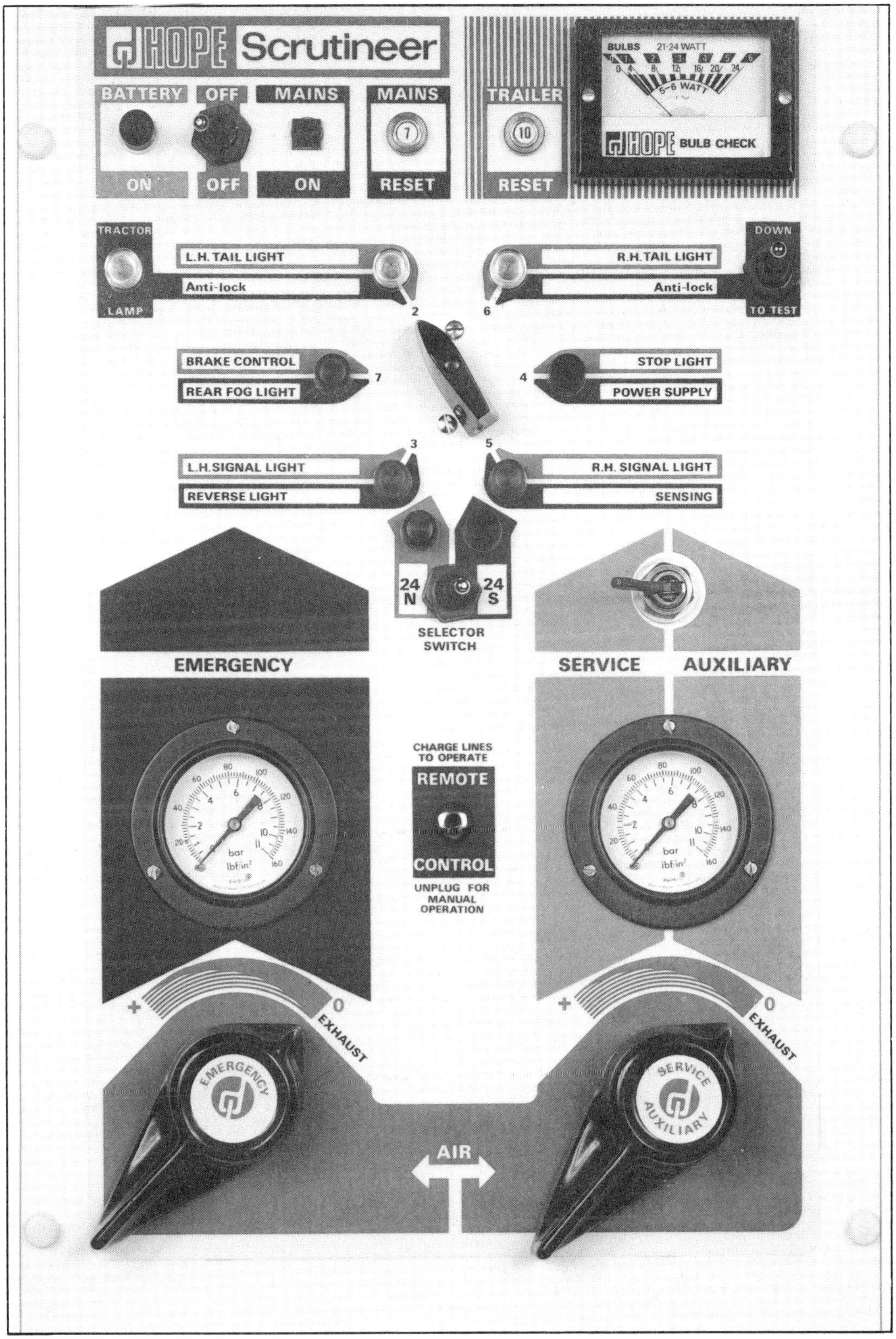

Figure 9.1 Hope Scrutineer

developed to enable the fitter to introduce air into the pneumatic braking system, without running the vehicle's engine. This has an extra attraction when the brakes on a trailer are being tested, as a tractor is not required just to supply air when it could be in use on the road.

Many fitters have developed pieces of equipment to do this job but there are several companies marketing a package that is delivered ready for use. One such company is Hope Technical Developments Ltd, which is already well known for its anti-jack-knife equipment. Their test equipment, called *Scrutineer* (see Figure 9.1), can be used to check for air leaks and for the application/release of valves, including anti-lock. In addition, it has an electrical aspect, as it can be used for circuit testing, battery testing and trickle charging.

This equipment has some interesting features. The air supply can be either taken off the workshop air circuit or from a compressed air (not oxygen) bottle, so it is readily available for breakdowns, as well as for routine work. It has a plug socket for a wander lead (battery voltage). The unit weighs about 40 lb (18 kg) and measures 23 × 14 × 8 in (580 × 380 × 200 mm), so it is even suitable to be carried in the boot of a car, providing only a small air bottle accompanies it.

Ideally, this equipment is used to test the brake equipment, and indeed the electrical equipment, of the much-abused semi-trailer, which is pulled by a variety of vehicles and drivers, but not treated with the same reverence as the prime mover.

Air can be introduced into the emergency line of a trailer by connecting it to the test equipment, instead of to the drawing unit. The relay emergency valve can be checked to see what pressures cause it to apply or release the trailer brakes, and whether it is sluggish in operation or in exhausting. Similarly, checks can be made for leaks. The service and secondary system can also be checked by the same method.

Test cube (BW)

This portable, suitcase-size device (see Figure 9.2) has been produced by Bendix Ltd, specifically for use with HGVs and PSVs, for checking and testing pneumatic components and/or systems, both on and off vehicles and trailers. It weighs 39 lb (17.5 kg), and measures 11.8 × 11.8 × 7.9 in (300 × 300 × 200 mm), making it easily portable and suitable for use either in the workshop or at the site of a breakdown. It can either be connected to a workshop airline or to a bottle of compressed air (*not* oxygen).

There are seven push-in couplings to receive hoses, which allow for a main supply from the air source, up to four inlets and two exhausts. Seven pressure gauges are included to monitor the various systems. Thus the test cube is capable of coping with all modern air brake equipment.

Figure 9.2 Test cube (BW)

Maintenance

All the leading manufacturers of air brake equipment lay down schedules of recommended maintenance in respect of valves and components. As there are variations in most of these units, it is impractical to list all types and, consequently, the maintenance schedules of only two companies (Bendix Ltd and Clayton Dewandre Co Ltd) are shown.

In general, most valves should be subjected to an operational and a leakage test at least once a year. However, there is no substitute for following the schedule in the vehicle workshop manual or that of the brake component manufacturer.

Sealing pipework connections can be achieved by the use of a specialist sealant. There are a number of these marketed and one such product, *System Seal*, contained in an easy-to-use dispenser, is shown in Figure 9.3.

Most manufacturers produce overhaul kits for the majority of their valves, but the exceptions vary with different companies, for example WABCO Ltd do not produce kits for system-protection valves. Most companies market exchange units. It is of the utmost importance to only use genuine overhaul kits or exchange units, as the failure of an anonymous part,

Figure 9.3 System seal (WO)

produced by an unknown company, may well cause a serious accident and jeopardise your job or the very existence of your company.

Note: All manufacturing companies are interested in preventing counterfeit components or units which affect their reputation, and welcome such items being brought to their attention.

WARNING

On the following pages there are frequent references to the use of suitable solvents. The utmost care should be taken in choosing solvents, bearing in mind the following points:

1 Fumes from most chemicals can be a hazard to health and/or a fire risk.
2 Dermatitis can result from allowing certain chemicals to come into contact with the skin. Therefore, protective gloves (for example, PVC) should be used to avoid this risk.
3 Some substances can injure aluminium castings which are the most common valve bodies in general use at present.
4 Rubber and synthetic materials should not be immersed in solvents.

One solvent that is widely used in industry is 1.1.1. trichloethane, which is marketed under various trade names, for example ICI use the trade name of *Genklene*. This substance is well known to HM Factory Inspectors and their advice or that of the manufacturer should be sought in any cases of doubt.

Always read the instructions on the solvent container and even with the safer varieties, always use them in a well-ventilated area.

Compressors

Compressors (BW)

At vehicle service intervals: Check for oil and air leaks, and water leaks in the case of water-cooled compressors.

If the compressor is slow to build up pressure, remove the cylinder head and valve plates. Check cylinder bores and valve assemblies, and remove excessive carbon.

Compressors (CD)

Daily (if self-lubricated): Check the oil level in the compressor sump and top-up as necessary. With this type, drain and flush the sump and refill it with fresh oil every 5000 miles.

Every three months/25 000 miles for single cylinder; or three months/10 000 miles for twin cylinder: Clean the air filter. Check security and drive. Check for leaks on the compressor and at unions. Check the torque on cylinder head bolts. Remove delivery valve caps and springs, and check for excessive carbon in head passages. Correct if necessary.

Annually/50 000 miles: Overhaul or replace the cylinder head.

Every two years/100 000 miles: The complete unit should be overhauled or replaced.

Unloader valves

Simplified valve (BW)

Monthly: Charge the vehicle's reservoir and check the cut-out pressure of the unloader valve. Correct, if necessary, by means of the screwed bush. Reduce the reservoir pressure and note the cut-in pressure. Recharge the system to cut-out pressure and stop the engine. Reduce the reservoir pressure to about 6 lbf/in^2 (41 kN/m^2, 0.4 bar) below cut-out pressure. Apply a soapy solution to the exhaust port and to the plastic cap on the control spring housing. Check for leaks. Any combined air leakage must not exceed a 1 in (25 mm) diameter soap bubble forming within 8 seconds. Reduce reservoir pressure to about 15 lbf/in^2 (100 kN/m^2, 1 bar) below cut-in pressure. Again using a soapy solution, test for air leaks around the exhaust port and the plastic cap on the control spring housing. If leaks are detected in any of these tests, which cannot be corrected by merely tightening connections, remove the unit from the vehicle, dismantle it, then inspect and replace defective parts.

CD line-type valve

This is sometimes called an unloader/governor valve.

Monthly: Check the security of the valve and connecting pipelines. Examine the exhaust diaphragm and replace it if necessary. Ensure that the vent in the top cover is not blocked.

Every three months: Inspect the air filter and renew as necessary (a new gasket may be required when the inlet cover is refitted). Check the functioning of the valve and for leaks, by the following method. Drain the reservoirs. Run the engine, at just below the maximum r.p.m. and check the length of time that the low-pressure warning buzzer continues to sound. If this time is excessive (3 minutes is allowed by the Department of Transport), check (if the compressor is not suspected) whether air is escaping from the unloader exhaust check valve or exhaust air line. The valve should be reconditioned or exchanged. If the vehicle reservoirs fail to hold pressure, causing the pressure to drop rapidly from cut-out to cut-in, the non-return valve in this valve is probably defective, and the valve should be overhauled or exchanged.

Check the pressure at which the valve cuts out using the cab gauge (service system), and check this against the vehicle's data in the workshop manual. In cases of doubt, a test gauge can be used to verify the accuracy of the cab gauge. Run the engine and gradually reduce reservoir pressure, by applying and releasing the foot valve. Note the pressure at which the unloader cuts in to allow recharging to commence. The pressure differential depends upon the setting of the particular valve and cannot be altered. This figure should be between 12–18 lbf/in^2 (83–124 kN/m^2, 0.8–1.2 bar).

To adjust this type of valve, slacken the locknut and turn the adjusting screw clockwise (to raise pressure) and vice versa, retightening the locknut upon completion of the adjustment.

Charge the braking system *to just below cut-out pressure* and then stop the engine. Cover the whole of the unloader valve connections and the outlet from the exhaust pipeline (if fitted) with a soapy solution. A tolerance allows for a soap bubble of up to just over $\frac{3}{4}$ in (20 mm) diameter forming within 4 seconds. Any excessive leakage indicates that one or more of the following items is defective: inlet valve seat; governor valve sealing ring; governor inlet valve; or unloader plunger sealing ring.

Leakage from the unloader exhaust check valve or port has the same soap bubble tolerance. Excessive leakage indicates that the unloader valve or seat are defective. No leakage is permissible from other parts of the valve or from connections.

Charge the system *until the valve unloads* and then stop the engine. Cover the governor exhaust check valve with soap solution to test the governor exhaust valve or seat for leakage. The maximum permitted size for the soap bubble that may form is just over $\frac{3}{4}$ in (20 mm) diameter within 4 seconds, and any excess of this figure indicates that the valve is in need of attention. Similarly, coat the unloader exhaust valve or port, with soap solution, to test the non-return valve and the unloader plunger sealing ring. The above tolerance applies to this area of the valve. The valve body and plugs should be covered with soap solution, but leakage from these items should not be tolerated. The low-pressure switch (if fitted) should also be completely coated with soap solution. Leakage from the cover or gasket shows that the O-ring diaphragm is defective.

When, as a result of the above tests, the unloader valve is found to be defective, this component can either by reconditioned with an overhaul kit or exchanged.

Governor valves

D2 (piston type) valves (BW)

Note: Should it be required, a water-proofing kit is available for this valve.

At vehicle service intervals: The governor should be checked for correct operation as follows:

1 Exhaust reservoirs.
2 Run the engine and check the build-up time, watching the reaction of pressure gauges.
3 Check that the compressor is unloaded at the correct cut-out pressure.
4 Use soap solution to check the exhaust for air leakage (this must not exceed a 1 in (25 mm) diameter soap bubble forming within 5 seconds).
5 With the engine running, reduce the reservoir pressure and check that the governor cuts in again at the recommended pressure.
6 Using soap solution check the inlet valve for air leakage. No other air leaks at governor joints or pipe connections are permitted.

If the governor fails the above tests, it should be removed from the vehicle and serviced.

Note: All bearing surfaces should be coated with grease (Bendix Ltd specify *K4-756-01*) when rebuilding this valve.

Diaphragm-type valves (CD)

Monthly: Check the security of the valve and connections. Check that the vent hole in the top cover is clear and that the lock nut on the adjusting screw is tight.

Annually/100 000 miles: Overhaul the unit or exchange it. The unit is checked as follows. Charge the system and note the pressure at which this valve cuts-out, correcting the adjustment if necessary. With the engine running, reduce the reservoir pressure gradually, noting the cut-in pressure. If the cut-in pressure is more than 25 lbf/in^2 (172 kN/m^2, 1.7 bar) below the cut-out pressure, the valve needs overhauling or replacing.

Leaks can be detected by covering the entire valve and connections with soap solution and slowly increasing the pressure, up to the cut-out point. If leakage occurs during this charging, from the exhaust diaphragm, the inlet seat or disc valve is faulty. However, if the leakage occurs after the cut-out pressure has been achieved, the exhaust seat, disc valve or sealing ring(s), is defective. Slight leakage, up to a bubble not exceeding approximately $\frac{3}{8}$ in (10 mm) in diameter can be tolerated. A leak from the vent in the cover, indicates that the diaphragm needs replacing. Leakages from other parts indicate that the valve requires attention.

Single-check valves (BW)

This component should periodically be checked for air leaks.

Non-return valves (CD)

These require periodic general leakage checks, using soap solution around connections. No leaks should be tolerated. Remove the inlet pipe from the non-return valve and apply soap solution around the valve. The maximum permissible leak would be indicated by a $\frac{3}{8}$ in (10 mm) diameter bubble forming within 5 seconds.

Safety valves (pressure-relief valve) (BW)

Use a soap solution to check for leaks. The maximum permissible leakage at the exhaust port would produce a 1 in (25 mm) diameter bubble forming within 4 seconds.

Safety valves (CD)

Daily: Listen for leaks.

Annually/100 000 miles: Replace or overhaul the valve. Set the valve to the figure given in the workshop manual data and test that the valve is working correctly. The normal setting is 160 lbf/in^2 (1103 kN/m^2, 11 bar).

DV-2 drain valves (BW)

At vehicle service intervals:

1 Clean the outside of the valve.
2 Fully charge the relevant storage reservoir.
3 Coat the valve and its outlet with soap solution. Note that no air leaks are permissible.
4 Test the action of the valve by making several applications and exhausting the relevant control valve (foot or hand), to reduce the air pressure contained in that reservoir. Only a limited discharge of air (and possibly condensate) should occur.

A defective valve should be removed from the reservoir and overhauled.

Note: A continual, slight leak could be due to excessive leakage elsewhere in the system.

Alcohol evaporator (BW)

Every three months/10 000 miles: Inspect the filter and evaporator tube to ensure that they are not restricted in any way. Check that the jar is free from sediment and clean if necessary.

Annually/50 000 miles: Completely dismantle and clean the parts in a suitable solvent. Automatically renew all gaskets. Inspect the body and jar for damage, replacing if necessary. Check the operation of the unit, when the compressor is charging the reservoirs. Bubbles should be present in the alcohol. If they are not visible, check all the connections and the gasket for tightness.

Alcohol evaporator check valves (BW)

Periodically check the operation of these valves and also check for leakage.

Alcohol injector (BW)

Weekly: Inspect for leaks and damage. Top up the reservoirs if necessary.

At vehicle service intervals: Drain and remove reservoirs. (Note that the alcohol vapour is poisonous and should not be inhaled.) Clean the filter. Remove the injector cap and verify that the plunger is moving freely whilst the governor is working. A soap solution should be applied to the filter when the valve is *in situ* and a watch kept for bubbles that may form. If bubbles appear when the valve is working, this suggests that the O-ring at the top of the pump plunger is defective. If bubbles appear when the compressor only is working, this suggests that the non-return valve in the unit is defective. A defective O-ring on the adjustment spindle will allow air to escape along that spindle. When refitting the reservoir to this type of unit, rotate the reservoir until the filler plug is now directly above this filter.

If the injector is dismantled, all the parts should be cleaned in a proprietary cleaning fluid, rinsed and thoroughly dried. All O-rings, seals and gaskets should automatically be renewed. Inspect all parts. Renew both springs, if damaged, or in the case of the plunger springs if the free length is less than $1\frac{15}{16}$ (49 mm). Smear all O-rings with grease (Bendix Ltd specify *K4-756-01*) before rebuilding the unit.

Condenser (CD)

Every three months/25 000 miles: Remove from the vehicle, taking care not to allow dirt to enter the ports of the auto-drain valve, if it is disconnected at this stage. Similarly, with different layouts, this pipe can be disconnected on the bench, but the same care should be taken to avoid dirt entering this unit, which is not being dismantled. Disconnect the auto-drain valve unit from the cylinder. Remove the filter from the condenser. If it is of paper construction, renew it with a copper mesh type. If a copper mesh type is fitted, clean it thoroughly. Clean out the cylinder. Inspect both sealing rings and replace if necessary. Smear the rings with *CDS 156* grease before refitting. The baseplate nuts should be tightened to 16 lbf/in (1.8 Nm).

Every six months/50 000 miles: Renew or overhaul the unit.

Checking correct functioning

To check the correct functioning of the unit and the auto-drain valve, build up the pressure in the system to the cut-out point of the unloader valve or the governor valve, whichever is fitted to the vehicle. In the case of the unloader system, the auto-drain valve should discharge at the cut-out pressure. However, with the governor system the air pressure must be reduced, by using the brake valve, to the cut-in pressure of the governor valve and at this point the auto-drain valve should discharge to atmosphere.

Air leaks

Systems with an unloader valve: Release all air pressure from the system. Remove the bridge pipe from the cylinder and fit a blanking plug. Run the engine. Cover the drain port with soap solution and watch for air leakage. A tolerance of a soap bubble of approximately $\frac{3}{16}$ in (5 mm) diameter forming within 5 seconds is permitted, but a leakage in excess of this amount indicates that the upper valve or seat is defective. Recoat the valve with soap solution and allow the pressures in the system to build up, watching the leakage. When the unloader valve cuts out, the bridge pipe can be reconnected. Reduce the pressure in the system, until the unloader valve cuts in and the compressor starts to recharge the system.

Coat the drain port and breather hole with soap solution. A slight leak from the breather hole indicates a defective piston sealing ring. Excessive leakage from this hole indicates a split lower diaphragm. A leakage from the drain port can be tolerated up to a $\frac{3}{8}$ in (10 mm) diameter bubble forming within 5 seconds. Fast leakage from the drain port indicates a split in the upper diaphragm.

Systems with a governor valve: Charge to cut-out pressure and stop the engine. Coat the breather and drain port with soap solution. A tolerance is allowed for leakage from the drain port, up to a soap bubble of $\frac{3}{8}$ in

(10 mm) diameter forming within 5 seconds. Leakage indicates that the lower valve or seat is defective. Excessive leakage from the drain port indicates a split upper diaphragm. This diaphragm defect could prevent the governor valve from causing the compressor to unload. A slight leakage from the breather hole indicates a defective piston sealing ring, whereas excessive leakage indicates a split in the lower diaphragm. The pressure should be reduced to cut-in pressure, so that the drain valve exhausts. Leakage from the drain port in excess of a $\frac{3}{16}$ in (5 mm) diameter soap bubble forming within 5 seconds suggests that the upper valve or seat is defective.

Low-pressure indicator (LP-3) (BW)

At vehicle service intervals:

1 Check the electrical connections and that the valve is functioning correctly.
2 Fully charge the reservoirs and check that no air leaks are present at the LP-3 air connection.
3 Stop the engine.
4 Check each LP-3 electrically.
5 Using control valves (foot and hand) watch the cab pressure gauges and note the pressure when the warning device operates. Recharge the reservoirs each time, before checking the next LP-3.
6 If the LP-3 is fitted as a stop light for spring brakes, check that these lights illuminate when the pressure drops.

Note: Service kits are not available, and a replacement unit is required when a defect cannot be corrected.

Stop-light switch (BW)

Periodically check for leakage and for correct operation. Ensure that the electrical connections are covered by a rubber boot.

Dual foot valve (Delta) (BW)

At vehicle service intervals:

1 Lubricate the treadle pins or operating linkage pivots, with a grade A grease (Bendix Ltd specify *K4-756-01*) or any good quality chassis grease.
2 Apply soap solution to this valve, especially around the exhaust ports. Fully apply the treadle and hold it down, testing for air leaks. Leakage should not exceed a 1 in (25 mm) diameter bubble forming within 5 seconds. When the foot valve is not being applied, leakage from the exhaust port must not exceed a $\frac{1}{2}$ in (12 mm) diameter bubble forming within 10 seconds.

Dual foot valve (concentric) (CD)

Daily: With the system fully charged, fully apply the treadle and watch the gauges. Note the pressure drop in 1 minute and if this is excessive, a further investigation should be made. Next, release the treadle slightly, and observe the gauges. The treadle should then be quickly released and note taken of any delay in releasing.

Monthly: Check the security of the valve and the condition of the pipework connections. The plunger should be lubricated. All treadle or linkage pivots should be lubricated with a good quality heavy oil. Gaiters should be inspected and renewed if necessary. The treadle linkage should be inspected for wear, excessive play, and any malfunction. The treadle stop should be adjusted to keep the slack in the linkage within the limits set by the manufacturer.

Annually/50 000 miles: Replace or overhaul the unit.

Caution: Clayton Dewandre Co Ltd issue the following warning in their service literature:

Due to widely different designs of dual concentric foot valve being manufactured, it cannot be overstressed that a proper understanding of the internal working parts be obtained by fitters who will be carrying out the procedures given under 'Dismantling' and 'Reassembly'. It must *not* be mistakenly assumed that knowledge acquired by dismantling and assembling one type will be applicable to all types. This is even more important when it is realised that several different stages of design improvement are manufactured concurrently.

As a result of the manufacturer's warning, it is suggested that the fitter has *at least* the assistance of the appropriate service literature from Clayton Dewandre Co Ltd (or the vehicle workshop manual) before tackling the overhaul of the units.

Air leakages

Charge the system to governor, or unloader cut-out point. Cover the exhaust port with soap solution. Leakage cannot be tolerated. Next, depress the treadle to a point where 90 lbf/in^2 (620 kN/m^2, 6 bar) is delivered to the brake chambers, and hold the treadle steady. Again coat the exhaust port with soap solution, with the treadle in this position. Leakage can be tolerated up to a $\frac{3}{8}$ in (10 mm) diameter soap bubble forming within 5 seconds. The brakes should then be fully applied and soap solution applied to the exhaust port and to the pipe connections to the valve. No leakage is permitted from the pipe connections or from the valve body, although there is a tolerance on leakage from the exhaust. This leakage must not exceed a $\frac{3}{4}$ in (20 mm) diameter soap bubble forming within 5 seconds.

Single-diaphragm brake chambers (BW)

Safety: Ensure that the road wheels are chocked before undertaking this task.

At vehicle service intervals: These units should be cleaned externally and checked for air leaks. Apply soap solution to the clamp ring vent holes in the non-pressure plate and at the push-rod aperture.

Fully apply and hold the relevant control valve (foot or hand) and watch for any soap bubbles that may appear. No leakage is permitted at either the vent holes or the push-rod aperture, but a slight frothing is acceptable at the clamp ring.

In the event of damage or leakage this unit should be overhauled or replaced.

Double-diaphragm brake chambers (BW)

Annually/50 000 miles: Remove from the vehicle, dismantle, clean and inspect the parts. Automatically replace diaphragms and spring. (Similar replacements should be carried out on other chambers fitted to that axle.) The unit should be rebuilt and refitted to the vehicle. Tests for air leaks should be made with soap solution.

Spring brake chambers (BW)

(See the dismantling procedure.)

Caution: Do *not* attempt to overhaul without the proper equipment because of the very high spring power trapped within the unit (see page 133).

Remote spring brake chamber (BW)

Safety: Chock the wheels first.

At vehicle service intervals:

1 Clean the exterior and check for air leaks by applying soap solution.
2 Apply soap solution to the breather and to the rubber boot.
3 Use the hand valve to release this brake, after which the combined leakage must not exceed a 1 in (25 mm) diameter soap bubble forming within 10 seconds.
4 A damaged chamber or one with severe leaks should be overhauled or replaced.

Caution: Do *not* attempt to overhaul without the proper equipment, because of the very high spring power trapped within the unit.

Hand control valve (TC-2) (BW)

Safety: Be sure that the road wheels are chocked.

At vehicle service intervals: With the reservoirs fully charged and the TC-2 valve handle in the release position apply soap solution to the exhaust port. No leakage is permissible. With the valve handle in the applied position, air leakage at the exhaust port must not exceed a 1 in (25 mm) diameter soap bubble forming within 5 seconds.

A damaged valve, or one with excessive leakage requires overhauling or exchanging.

Spring-brake hand control valve (S1) (truck) (BW)

Safety: Be sure that the road wheels are chocked.

At vehicle service intervals: With the reservoirs fully charged, move the hand lever towards secondary/park, and check that the line pressure escapes to atmosphere whilst this is taking place.

With the lever at park, apply soap solution to the valve body and exhaust port. No leakage is permitted. If leakage occurs it will be due to a defective inlet valve.

With the lever in the 'off' position apply soap solution to the valve body and exhaust port. No leakage is permitted from the valve body, and leakage from the exhaust port must not exceed a 1 in (25 mm) diameter soap bubble forming within 5 seconds.

If this valve is defective or has a severe leak it should be overhauled or exchanged.

Spring-brake hand valve (L4) (BW)

Safety: Be sure that the road wheels are chocked.

At vehicle service intervals: Although this valve is being checked for damage, correct operation and air leaks, it is also extremely important that the condition of the handle shaft should be examined with great care. The manufacturers, in their service literature, make the following statement:

If there is the slightest suspicion that the handle shaft is bent the complete hand valve must be replaced. *Note*: Hand control valves have a limited life and *must* be replaced at three years, or before if necessary.

Fully charge the reservoirs. Fit a test gauge to a spring brake test point and fit another on the trailer secondary coupling head.

Move the handle to sec. Delivery output to the trailer secondary circuit, i.e. the blue line in three-line circuits and the yellow line in two-line circuits, *must* be at least 95 lbf/in^2 (655 kN/m^2, 6.6 bar). At the same time pressure to the tractor spring brakes must drop to zero.

Move the handle to the 'off' position, and the air to trailer brake coupling head should release, whilst the pressure to the tractor spring brakes should rise to a minimum of 95 lbf/in^2 (655 kN/m^2, 6.6 bar).

Move the handle to park, and zero pressure should exist at both the secondary coupling head and at the spring brakes of the drawing unit.

Leakage test: With the valve handle in the 'off' position and soap solution applied to the body and exhaust port, no leakage is permitted from the body. Leakage from the exhaust port must not exceed a $\frac{1}{2}$ in (13 mm) diameter soap bubble forming within 5 seconds.

With the handle in the secondary position, no leakage is permitted.

If the pressures quoted are not achieved, or the leakage rate quoted is exceeded, the valve should be overhauled or exchanged.

Differential-protection valve (BW)

At vehicle overhaul periods: Remove the valve from the vehicle, dismantle and wash the parts in a suitable solvent, then examine the parts for wear and damage. Renew all O-rings and circlips, as well as unserviceable parts. Lubricate the new O-rings with grease (Bendix Ltd specify *K4-756-01*). Rebuild the unit.

Apply pressure to the service and secondary inlet ports, in turn, keeping the two delivery ports plugged during the test, while using soap solution to show up any leaks on this valve. Refit the valve to the vehicle, check its operation and for leaks.

Spring brakes

The very name 'spring brakes' means, to some, the ultimate in safety. The assumption is that they will stop you dead if the air pressure fails. This is only a half-truth, as many people have been surprised to discover.

They are an excellent device, but they still require attention and maintenance. A simple example is that of a perfectly efficient set of spring-brake chambers which can be rendered ineffective owing to the lack of adjustment of slack adjusters. A prohibition notice PG9 (previously GV9), the jeopardising of an operator's licence, involvement in an accident - all or any of these can occur owing to such a simple omission in a maintenance scheme. Naturally, spring-brake chambers should be overhauled and/or replaced at manufacturers' recommended service periods, for example annually. Similarly, leakage tests should be conducted when maintenance is scheduled.

Another point is that with the wide range of different spring brakes being fitted onto vehicles, it is important that certain dangers are appreciated. Even the simple task of recovering a vehicle which has spring brakes fitted, when the air pressure has failed, can have greatly varying procedures, and includes the following:

1 The wind-off bolt, at rear (see Figure 6.77).
2 The wind-off collar, at front (see Figure 6.71).
3 No device (when the slack adjuster or other mechanical adjustment must be undone) (see Figure 6.72).
4 A quarter-turn rear wind-off arrangement (see Figure 9.4).
5 Knob at rear which is struck a single blow to release the brakes (see Figure 9.5).

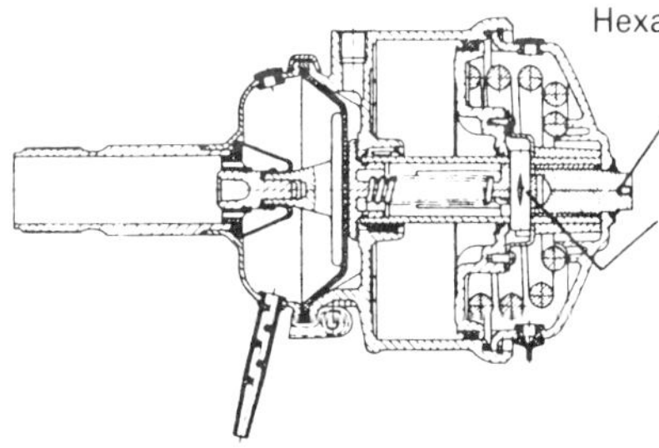

Figure 9.4 Quarter-turn wind-off

Whichever type is in use, always chock the road wheels *first*, as once the spring brakes are freed the vehicle may roll away, and this has happened.

Servicing a spring brake

Assuming that all is well and the brake chamber has been removed and put on to the workbench:

1 Make sure that the chamber is *correctly identified*.
2 Make sure that the *correct dismantling procedure* for that chamber is *known*.
3 Make sure that the *correct tools/equipment* are available.

In most cases a mechanical press is the major piece of equipment required because its slow action is easier to

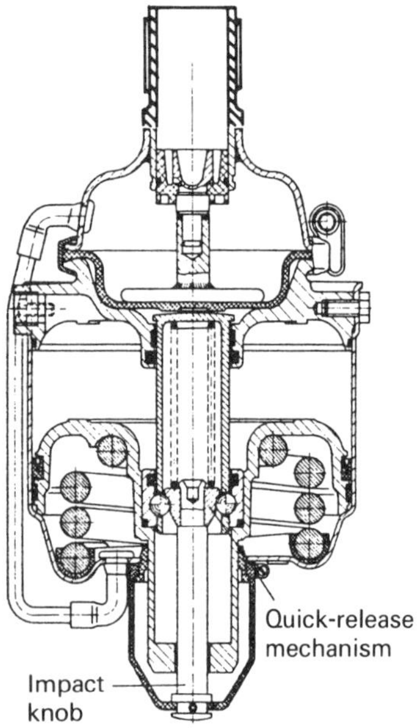

Figure 9.5 A single-blow release

control than that of the less-safe mechanical press.

Remember: The spring that is compressed within the brake chamber is as powerful as a car suspension coil spring.

Figures 9.6 and 9.7 show two totally different methods necessary, even with a press.

Reconditioning procedure for a Clayton Dewandre pull-type spring-brake chamber (without wind-off)

Safety: Chock the road wheels first.

Removal from the vehicle: General guidance only can be given here and it is suggested that the workshop manual is consulted.

Usually the procedure is to charge the braking system to cut-out pressure of the unloader valve or governor valve and then move the hand control lever to the 'off' position in the gate. If it is found that the pull-rod is one piece from the unit to the clevis, this will indicate that the chamber is *not* a wind-off type. The engine should be left running, and the hand control in the 'off' position. Mark the pull-rod 2 in (50 mm) from the body of the chamber.

Remove the clevis pin whilst the spring is held by the air. The hand valve lever should be moved to park and a note made of the position of the mark that was made on the pull-rod. This mark should be approximately level with the body and if the mark is not reaching the body this indicates that either the piston or the interior of the body are defective. The unit should then be unbolted from the vehicle for stripping/inspection. A note should always be made of the length of the pull-rod or thread into the clevis. The cylinder and body should be marked to aid reassembly, after stripping.

Stripping: As shown in Figure 9.6, place the unit into the special press and secure it to the base by using the nuts on the mounting studs. Select the correct adapter and then tighten it down sufficiently to hold the unit. Remove the setscrews positioned around the circumference of the cylinder. Unwind the centre rod of the main tool and the spring should extend in length, until it is fully extended, thus freeing the piston and body. A seizure can be released by returning the centre stem of

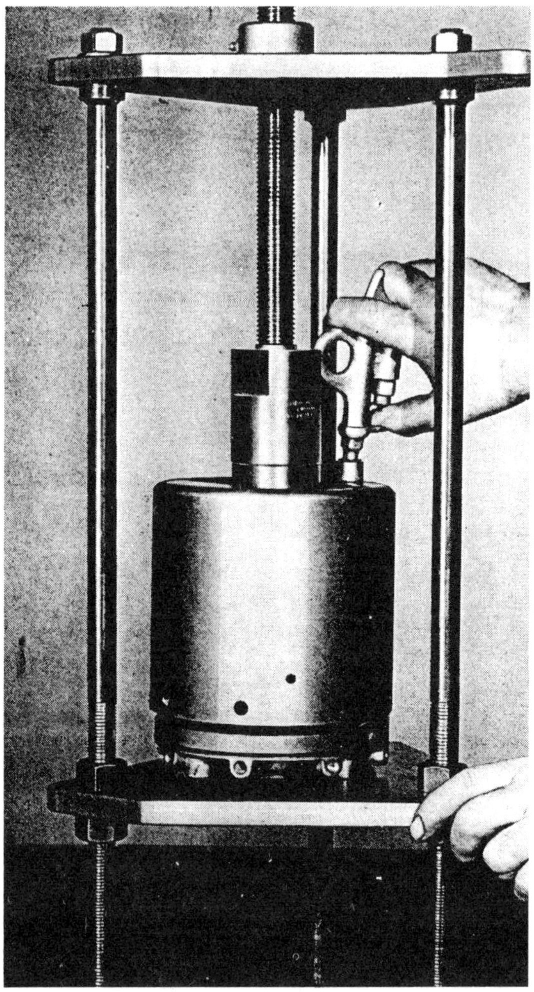

Figure 9.6 Using compressed air to aid the dismantling of a remote spring-brake chamber whilst the pressure is being slackened off, until the seal is expelled from the cylinder

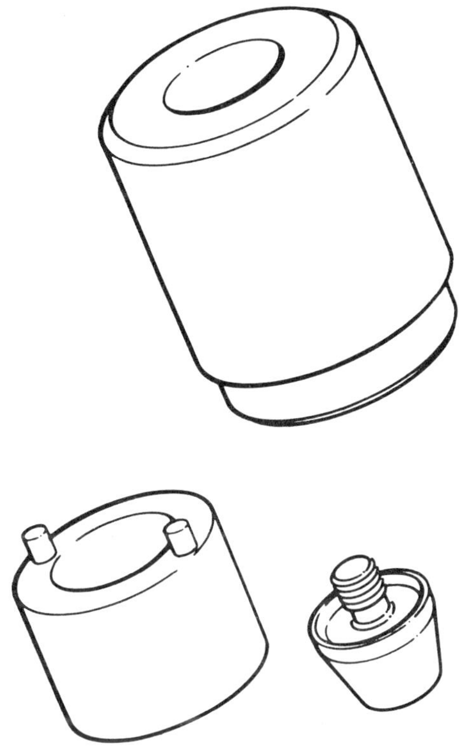

Figure 9.7 The three special adaptors necessary for dismantling a Bendix spring-brake chamber

the main tool, close to the adapter. Compressed air can then be applied to the port in the cylinder, but care must be taken that the adapter and centre stem are very close, as when the components suddenly part, the spring force is very high. Obviously all damaged parts should be replaced and an evaluation made of any internal damage.

Reconditioning: All parts should be cleaned in a suitable solvent and then dried with an air-line. A general inspection should be made for cracks, dents and general damage, including damaged threads. The usual major overhaul period is 50 000 miles or annually, when a repair kit should be used automatically.

Rebuilding: The sliding surfaces of the cylinder, piston, piston shaft, all load-bearing surfaces and the whole of the spring, should all be treated with a suitable grease. The felt lubricator wick should be soaked for 30 minutes in suitable oil. Reconnect the pull-rod to the piston shaft. Replace sealing rings, wick, piston and springs, line-up the dismantling marks, then compress the spring. Replace the setscrew and tighten to a torque of 8 lbf/ft (11 Nm).

Check with the workshop manual as there are variations.

Reconditioning a Bendix spring brake

As shown in Figure 6.73 on page 55, this conventional spring-brake chamber consists of two separate chambers. Note that there are remote spring-brake chambers in use which only have the spring chamber and are used solely for secondary/park brakes.

The two separate chambers are:

1 a conventional single-diaphragm brake chamber;
2 a spring-piston brake chamber.

The latter contains a very powerful spring which can cause grave injuries if unleashed by removing the anti-explosion washer and the large circlip. Therefore, the dismantling of the spring side of the brake chamber should *not* be undertaken unless special equipment and tools, such as the *Churchill* rig/tools, are available. The workshop manual should be consulted. The following is an example of the methods employed.

Removing the unit: To remove the complete unit from the vehicle, exhaust the air from the secondary system. If the secondary section of the unit is functioning correctly the brakes will then be applied. Chock the road wheels. Wind-off the release bolt or other device fitted, to release the brakes. Disconnect the clevis pin from the yoke. Disconnect both pipelines from the chamber. Now unbolt the unit from its mounting.

On the bench: After cleaning the exterior of the component, remove the nylon breather tube. Remove the yoke from the operating rod. Fit a piece of steel tubing over the operating rod, with a large washer at either end to take the thrust. Tighten a nut down the thread, against the tube, thus compressing the ordinary spring in the service-brake chamber. Undo the bolts securing the clamp ring and remove it. Lift out the operating rod, return the spring and non-pressure plate, as one unit. Also remove the diaphragm. If this section of the brake is to be dismantled, hold the operating rod, in soft clamps, in a vice and undo the nut.

Spring chamber: Using felt or soft material, grip the push-rod, at the front of the chamber using mole grips (or a similar device) and then unscrew the setscrew that holds the push-plate to the push-rod. (It may be necessary to ease the push-plate out a little, with a screwdriver, before this operation is started.)

Next, use the special tool (cylindrical in shape, with

two pins as in Figure 9.7), together with a $\frac{1}{2}$ in (12 mm) drive speed brace, to remove the anti-explosion washer from inside the unit, at the small end of the chamber.

Place the chamber on the base of the main tool, within the three legs. Ensure that the central stem thread is well lubricated with hypoid oil, without contaminating the brake unit. Using the plain, stepped, cylindrical tool (see Figure 9.7), placed in the narrow bore of the chamber, wind down the centre shaft of the main tool, having lubricated the spigot, until contact is made. Scrape away any sealing compound from around the circlip. Remove the circlip from its groove, then carefully unwind the centre screw of the rig, to release the spring pressure.

Service: All metal parts should be cleaned in a solvent. All parts should be examined for excessive wear or damage. If body or head bore wear exceeds 0.010 in (0.25 mm) the item should be replaced.

Before rebuilding, all seals and mating surfaces should be lubricated according to the manufacturer's recommendations.

Rebuilding the unit: Fit a new seal and a new wiper to the piston, applying grease to their grooves. Felt wipers need soaking in recommended oil.

Fit two new O-rings to the collar, then fit the collar to the push-rod, using a new circlip. A new seal should be fitted to the push-rod.

Refit the push-rod into the piston with a new circlip, screw the taper adapter (using the special tool - see Figure 9.7), only finger tight, into the end of the push-rod and lubricate.

Refit the piston, at an angle, into the chamber. Push the piston to the bottom of the chamber, so that the special tool enters the seal without causing damage. The special tool can now be removed.

Lubricate and refit thrust washers, at each end of the main spring. Refit the end of the chamber by using the plain cylindrical tool in the narrow diameter of the end plate and the pressure of the main rig. Make sure that the spring, thrust washers and end cap, are seated correctly.

Refit the circlip or preferably fit a new circlip. Check that it is *correctly* fitted before releasing the press.

Lubricate the bore of the head and fit a new anti-explosion washer.

A sealing compound, as recommended by the manufacturer, should be applied around the circlip and allowed to dry. Refit pressure plate/setscrew.

Reassembly of a non-hazardous chamber: This consists of refitting the items as one unit and then fitting the clamp ring. Torque figures are quoted, with a recommendation that the torque should be rechecked after an hour.

The warning tags should be refitted, together with locknuts. Finally, the release bolt should be fitted and fully tightened to 35 lbf/ft (47 Nm) torque.

Dealing with a double-diaphragm (piggy-back) spring brake

Note: This job can be done without even removing the spring brake chamber from the vehicle.

The overhaul of a piggy-back spring brake is shown in Figure 9.8. The steps to be taken are as follows:

1 Stand the vehicle on level ground, with the wheels chocked and set the spring-brake control to park.
2 Remove the release tool from its sleeve. Clean and oil the bolt.
3 Remove the rubber plug from the rear of the chamber. Fit the T-head of the special bolt into the T-slot in the pressure plate (by feeling it is correctly positioned) and turn it a quarter-turn in a clockwise direction to lock the T-head into the pressure plate.
4 Put the washer and nut on to the release bolt, then tighten down, winding off the spring.
5 When the nut is tightened, the split clamp ring can be removed with safety.
6 The complete spring chamber can then be slid away from the service chambers complete with the diaphragm.

Reassembly:

7 A new diaphragm should be slid into position, then the chamber assembly can be replaced (assuming that the assembly does not also require replacement). The clamp ring can be replaced.
8 In the cab, the spring-brake control can be moved to the 'off' position. Check for air leaks. Light tapping of the split clamp ring, plus retightening the clamp bolts should stop any leaks from that area.
9 Now the release bolt, nut and washer can be undone.
10 By giving a quarter-turn anti-clockwise the bolt, nut and washer can be removed and then replaced in the sleeve.
11 The protective rubber plug can be replaced in the rear of the brake chamber.
12 Wheel chocks can now be removed.

In general

Do *not* accept any repair kits or exchange spring

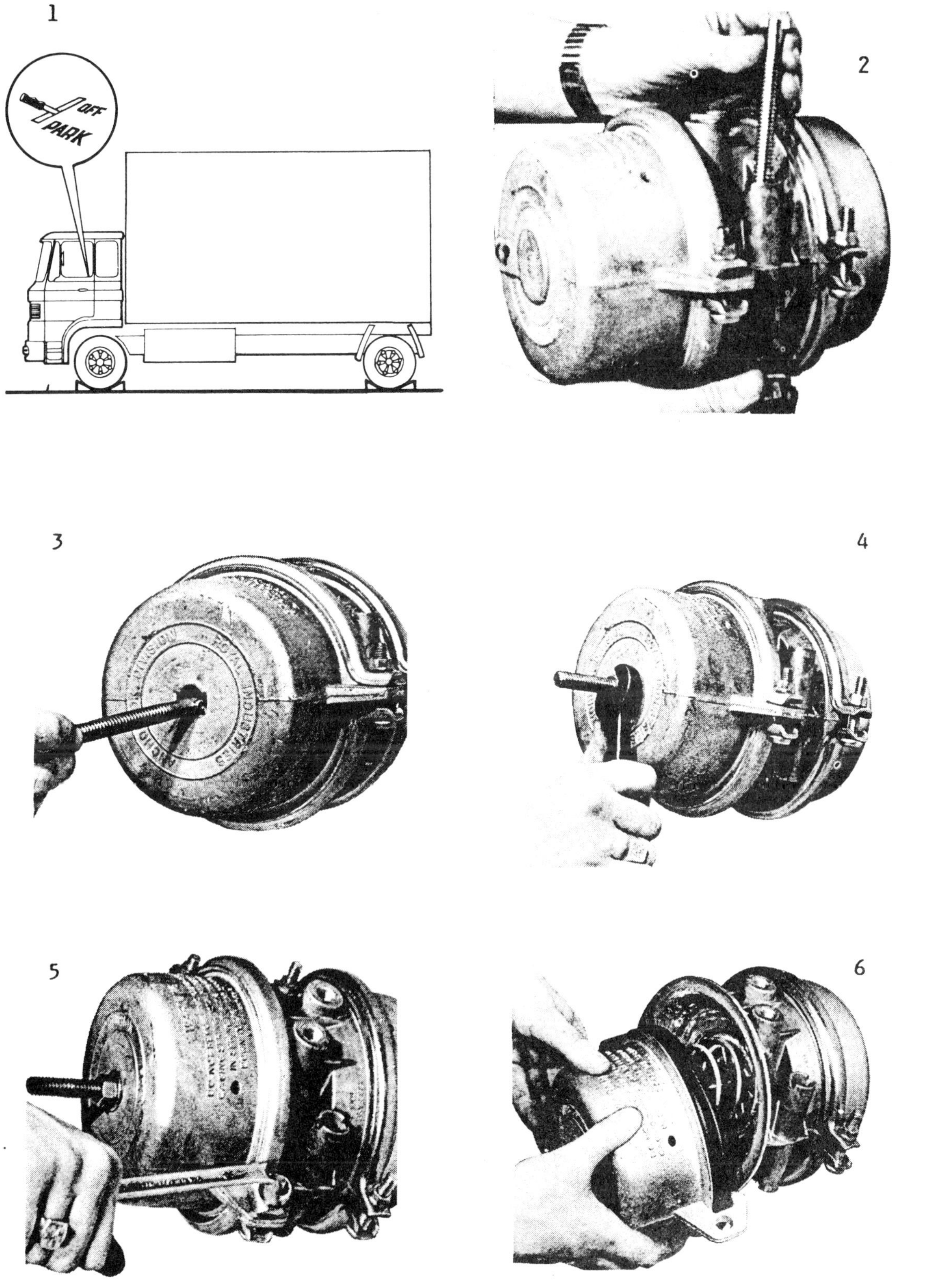

Figure 9.8 Overhaul of piggy-back spring brake

brakes, other than those belonging to a recognised brake manufacturer.

Relay valve (R6) (BW)

Every three months/10 000 miles: Check the relay valve for correct operation, as follows:

1 Check that the valve reacts immediately to changes in the signal line pressure and that pressure delivered by the relay valve is proportional to that received by its signal port.
2 Use soapy water to check that neither the valve nor pipe joints leak, whether the valve is in use or not.
3 In the event of the relay valve not reaching the standard set, it should be overhauled or replaced.

Relay emergency valve (RE-6) (BW)

Every three months/10 000 miles: Check the relay valve for correct operation, as follows:

1 Connect the tractor (or test equipment, with compressed air facility) to the trailer and exhaust the air from both systems.
2 Build-up the air pressure in both systems. Check that the trailer brakes apply during initial charging, and that they fully release when the emergency (supply) line pressure reaches 60-65 lbf/in^2 (414-448 kN/m^2, 4-4.5 bar). Check service brake application and release, together with the response at trailer wheels.
3 Charge the system to cut-out pressure and then switch off the engine. Cover the valve, valve exhaust port and pipe joints with soap solution and check for leaks. A leak from the exhaust port should not exceed a soap bubble of 1 in (25 mm) in diameter forming within 5 seconds. No other leakage is permissible.
4 Apply and hold the service brake on, and stop the engine. Again apply the soap solution test.
5 Disconnect the emergency (supply) line, and note whether the trailer brakes apply immediately. Reconnect and check that the trailer brakes release immediately.
6 With the engine switched off, use the foot valve to reduce the pressure from the trailer reservoir, and note when the trailer brakes automatically apply. This should occur when the trailer reservoir pressure is approximately 30 lbf/in^2 (207 kN/m^2, 2 bar), or at the pressure setting of the tractor protection equipment, if fitted.
7 If this valve does not pass these tests, it should be overhauled or exchanged.

Quick-release valve (BW)

Every three months/10 000 miles: The quick-release valve should be checked as follows:

1 Apply and release brakes and check that the air pressure rapidly escapes to atmosphere through the exhaust port of this valve.
2 Apply soap solution to this valve and its exhaust port. Apply the brakes. No leakage at the exhaust port in excess of a 1 in (25 mm) diameter soap bubble forming within 1 second is permitted. No leakage is permissible from any other part of this valve.
3 If this valve does not comply with these test requirements, it should be serviced or exchanged.

Variable-load valve (BW)

On new vehicles the road springs settle down and when this happens the vertical linkage that is connected to this valve should be corrected so that the braking force, for the unladen vehicle, is correct. Even when this linkage is out of adjustment because of this spring settlement, the laden vehicle will still receive maximum braking. The horizontal lever length is not affected by settlement and should not be adjusted. Periodically a check should be made for spring settlement and for leaks.

At service intervals: Apply soap solution to the valve body and exhaust ports. Make and hold a full service brake application. Air leakage from the exhaust ports must not exceed a 1 in (25 mm) diameter soap bubble forming within 10 seconds. Leaks from the body are not permissible.

Check the linkage between valve operating lever and vehicle axle(s) for condition and security.

Check the valve in both the fully-laden and unladen condition. The reservoirs should be at system pressure of 120 lbf/in^2 (827 kN/m^2, 8.3 bar).

If this valve does not comply with these test requirements it should be serviced or exchanged.

Power-assisted handbrake valve (CD)

Monthly: Clean off external deposits of abrasive matter and contaminated grease. Check the security of the valves and the unions. Check pivots, ratchets and pawls for wear, damage and mechanical efficiency. Diaphragms or filters that are fitted in ports, and found to be damaged or blocked, must be renewed.

Annually/100 000 miles: Replace or overhaul the unit.

As there are many types of valves and layouts for the air-assisted handbrake, the literature produced by either the vehicle manufacturer or the valve manufacturer, should be consulted when considering the repair of such units.

Slack adjusters (BW)

At service intervals: The slack adjuster should be inspected for wear or damage. Check that the brake actuator push-rod, yoke pin and lock nut are secure. Pack the unit with grease (Bendix Ltd specify *K4-756-01*).

Pressure-protection valve PR-2 (BW)

Periodically check for leaks. Note that a leak at the exhaust vent at the base of the plastic cap indicates that the O-ring on the piston or the piston plug is defective.

Every six months: Using a test gauge, check the pressure at which the valve closes when the delivery line pressure is allowed to fall. If this pressure exceeds a tolerance of approximately 5 lbf/in^2 (34 kN/m^2, 0.3 bar) from the recommended setting, use the adjusting cap to correct the situation, then retighten the lock-nut. In addition, a lockwire and seal can be fitted to prevent interference with settings.

Pressure-reducing valve RV1 (BW)

Periodically check for leaks.

Every six months: Check the delivered pressure leaving the delivery port. If this pressure exceeds a tolerance of approximately 5 lbf/in^2 (34 kN/m^2, 0.3 bar), this should be rectified by the use of the adjusting cap, at the top of the unit, after which the locknut should be retightened. A lockwire and seal can also be fitted to prevent interference with this component.

Pressure regulator and diverter valves (CD and BW)

Monthly/2000 miles: Check for security and leaks. Check that the breather hole in the cap is clear. Note the cab gauge reading when the auxiliary system starts to charge and correct the adjustment if necessary.

Every six months/10 000 miles (BW) With a test gauge, check the pressure at which the valve closes and if the setting is outside a tolerance of approximately 5 lbf/in^2 (34 kN/m^2, 0.3 bar) adjust the setting.

Annually/50 000 miles (BW); annually/100 000 miles (CD): Overhaul or replace the unit.

To overhaul the unit, remove it from the vehicle, dismantle it, clean the parts in a suitable solvent and examine for wear or damage. Replace the diaphragm and also replace any other parts as necessary.

Rebuild the unit and repeat monthly and six-monthly checks.

Supply pressure at about 10 lbf/in^2 (69 kN/m^2, 0.7 bar) below the setting of the valve. Cover the outlet port with soap solution and check for leaks. The permitted tolerance is a 1 in (25 mm) diameter soap bubble forming within 4 seconds for the Bendix valve, whereas the Clayton Dewandre valve tolerance is a 0.4 in (10 mm) diameter bubble in 3 seconds.

Double-check valve (BW)

Periodically check for leaks. No overhaul kits are produced by Bendix Ltd and, therefore, if a defect occurs a replacement valve is necessary.

Air drier (BW)

Periodically check for leaks and that the unit is secure. Check that the purge valve operates correctly whenever the compressor is unloaded by the governor.

At vehicle service intervals: Disconnect the heater wire and the air supply to the purge valve.

Remove the bottom casting from the unit so that the complete desiccant cartridge can also be removed.

Renew the oil filter if necessary. If the desiccant pellets have become contaminated, they too should be replaced. The body and filter plates should be cleaned. Inspect the desiccant plate and verify that the check valve is seated correctly and that the purge vent is clear. Rebuild the complete cartridge, but before replacing the bottom casting, check the purge valve for wear or damage. The valve spring, seat and O-ring can be replaced, if necessary. Check that the heater thermostat is working correctly and that the element is not clogged. Grease the large O-rings before assembly. Remove the safety valve from this unit and check that it discharges correctly at 150 lbf/in^2 (1034 kN/m^2, 10 bar).

Supply-dump valve (BW)

Monthly: Couple the drawing vehicle to the trailer and make sure that the reservoirs are then charged. Next disconnect the service coupling to the trailer, but leave any other couplings connected.

Make a *full* application of the foot valve treadle. Air should exhaust from the supply dump valve, showing that it is operating correctly. (Where self-sealing couplings are used, a dummy adapter is required to keep it open to atmosphere.) The relay emergency valve should also be seen to operate the brake linkage, when the foot valve is applied.

Check for leaks and damage.

Annually/50 000 miles: Remove the unit from the vehicle and carry out the following tests:

1 Plug three of the four ports that can be used to connect the valve to the emergency line and the relay emergency valve, and connect the fourth port to an air reservoir, not exceeding 61 in^3 (1000 cm^3) capacity, which is fitted with a pressure gauge.
2 Connect an air supply of 90 lbf/in^2 (620 kN/m^2, 6 bar) in the lowest of the three ports that are in a vertical line, i.e. the port that is connected to the trailer reservoir on the drawing vehicle. Check for leaks from the other two ports positioned vertically above this port, i.e. the ports that connect with the service line for the drawing vehicle and the service line for the trailer. Also check for leaks at the exhaust port, i.e. at the base of the unit. Excessive leakage through the centre of the inlet/exhaust valve is likely to be the result of a defective valve seat.
3 Apply an equal pressure of 90 lbf/in^2 (620 kN/m^2, 6 bar) to the top two of the ports that are in the vertical line, i.e. the sensing lines from both the trailer and the drawing vehicle service lines. The reservoir pressure should not be affected, as this is the normal driving situation, but if an excessive leak occurs at the exhaust port, defective O-rings on the piston are a likely cause of this fault.
4 Leave the top port (normally connected to the trailer service line), open to atmosphere and apply the pressure, already described, to the centre port (normally connected to the service line of the drawing vehicle), also leave the bottom port, beneath the other two, connected to a reservoir, as described in 2 above. Another reservoir is left connected, as described in 1 above.

   The latter reservoir should discharge to atmosphere, through the exhaust valve, but no other discharges should occur. If discharge through the exhaust does not occur, a jammed piston or a blocked filter are likely causes.

When dismantling this type of valve, clean all the parts in a suitable solvent and inspect for wear or damage. Automatically renew all O-rings, circlips, inlet/exhaust valve elements, the small valve spring, filter and exhaust flap. The piston spring should be renewed if its free length is less than 3.09 in (78.5 mm). When rebuilding this unit grease all O-rings with grade A grease (Bendix Ltd specify *K4-756-01*).

Trailer-control valve (BW)

Weekly: Check for damage and air leaks, apply soap solution to the valve body and exhaust flap, make and hold a full pressure brake application. Air leaking from the exhaust flap must not exceed a 1 in (25 mm) diameter soap bubble forming within 1 minute.

Monthly: Check for damage and air leaks as in the weekly maintenance procedure.

Safety: Chock road wheels.

Check for correct operation as follows:

1 Fully charge air reservoirs.
2 Isolate each service signal port in turn, i.e. 4.1 and 4.2. Apply the footvalve treadle and check that the drawing vehicle and trailer brakes respond rapidly when either section is tested. Sluggish brake operation indicates that a piston within the trailer-control valve is sticking.
3 With the engine switched off and the footvalve released, move the handvalve to the secondary position, and check that the trailer brakes apply. If the trailer brakes fail to apply, this suggests that a piston within the trailer-control valve is sticking.
4 Leakage at the exhaust flap indicates either a damaged inlet/exhaust valve or defective O-rings in that assembly.
5 If the trailer-control valve has a supply dump function, apply the following test: fit an open-ended pipe to the trailer service (control) coupling and fully apply the service brake treadle; then check that air escapes from the open pipe and that the trailer brakes are fully applied.

Note: Any operating faults or air leaks must be rectified immediately. Overhaul kits and instructions are available from the manufacturer, but great care should be taken when identifying the particular model and when overhauling such valves.

Anti-lock and traction control

In the event of a fault developing in either an ABS or an ASR system, it is essential that simple to use, robust, diagnostic equipment is readily available.

Terminology from the electronics and computer industries are encountered in the instruction books, but the test equipment is easy to use and to understand. Naturally, systems vary, and it is important that the particular checking system is fully understood.

Manufacturers run short courses to explain their own equipment and methods to workshop personnel.

Test equipment available from system manufacturers could include:

1 A diagnostic controller, which plugs into the diagnostic plug on the vehicle. This has a card slot, to receive a *smart card*, i.e. the appropriate card for the particular system. There are a series of push-buttons and a selection of tests.
2 An integrated multimeter which can either be used in conjunction with the test programme or in completely separate tests.
3 A liquid crystal display (LCD) displaying four lines, with 40 characters.

The above test equipment can even be connected to a printer or a personal computer.

Another system uses a hand-held diagnostic display unit that contains a LCD and a microprocessor, which when connected to the ABS system via a multipin connector, displays numbers to indicate faults. The operator then consults a simple handbook containing the list numbers and the faults that have occurred.

There are variations to both of the above systems, and indeed other systems are available, but the intention here is to indicate that all manufacturers are catering for simple diagnosis of faults in vehicles in service.

Conversion of two/three-line tractors and trailers

The conventional three-line (UK) tractor has three brake hoses to connect to its trailer:

1 the yellow service (control) line,
2 the red emergency (supply) line,
3 the blue secondary (auxiliary) line.

Likewise, the three line (UK) trailer has three corresponding connections.

The EEC-type, two-line tractor has two hoses:

1 the yellow service (control) line,
2 the red emergency (supply) line.

The two-line trailer has two suitable connections for these lines.

There are various methods of adapting the connections of the three-line tractor to enable it to tow, and satisfactorily brake, a two-line trailer. Figures 9.9 and 9.10 both show safe methods of conversion. However, these are not the only possibilities, although care must be taken to avoid the risk of air loss if a particular component fails. For example, the simple device of feeding both the service and the secondary lines into opposing ports of a double-check valve (change-over valve), does not provide the same safeguards as the use of a multi-relay valve.

If a two-line trailer is to be drawn by a standard (i.e. unconverted) three-line tractor, that trailer will need an additional connection. Figures 9.11 and 9.12 show two different approaches to this problem, but again these are not the only methods. Conversely, some methods are unsafe and again the simple use of a double-check valve (change-over valve) instead of a relay valve is potentially dangerous.

If a two-line tractor needs to draw a three-line trailer the service and secondary brakes on that trailer will work correctly in response to the driver's application of the respective control valve. However, this is still a grey area of UK law as the secondary diaphragms of the double-diaphragm brake chambers are not available for use in these circumstances. If such operations are contemplated, it is advisable to seek authoritative advice and obtain written confirmation of any such dispensation.

With any trailer that is fitted with anti-lock (electronic) brakes, there must be a suitable connection to the tractor's electrical supply. The conventional method is by the use of an ISO seven-pin connector (ISO 3731). Such conversions should be verified with the manufacturer of the anti-lock braking equipment.

Figure 9.9: This shows a three-line tractor that has been modified by the addition of a dual-relay valve (A), a CA-type male coupling (E) and a dummy coupling (F), complete with retaining chain.

For three-line operation the tractor and trailer operate as normal, but the new CA-type male coupling (E) is joined to a new dummy coupling (F), allowing this line to be vented to atmosphere, as for this operation it is surplus to requirements. For drawing a two-line trailer, it is necessary to join the old CA-type and the new CA-type couplings. Thus, only the two remaining hoses link the tractor and trailer, and when either a service or a secondary-brake application is made, compressed air travels to the trailer relay emergency valve via the yellow service (control) line.

Figure 9.10: This shows a three-line tractor that has been modified by the addition of a relay valve (B), a change-over (non-return) valve (D), a CA-type male coupling (E) and a dummy coupling (F), complete with retaining chain.

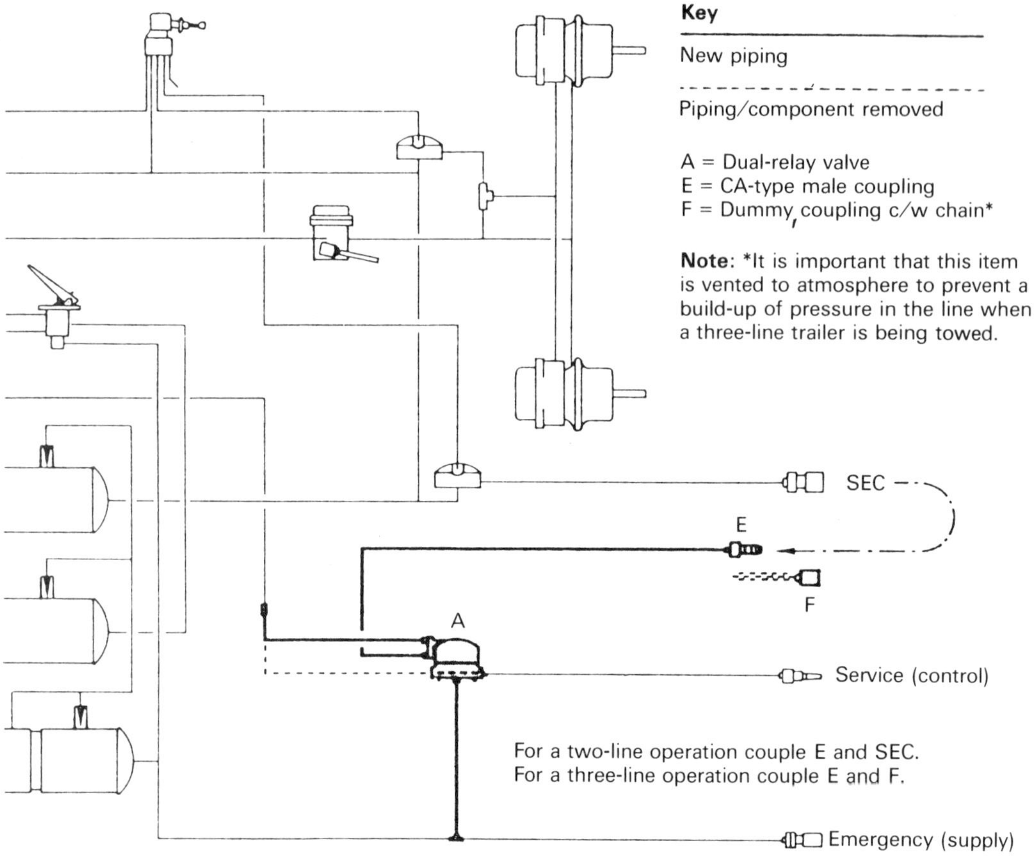

Figure 9.9 Tractor two/three-line conversion (1)

For three-line operation the new CA-type coupling is connected to the new dummy coupling, leaving the conventional three-lines connected. When a two-line trailer is to be drawn the conventional blue secondary (auxiliary) line is connected to the new CA-type coupling. Thus, when either the foot or hand control are used, compressed air travels to the relay emergency valve on the trailer via the yellow service (control) line.

Figure 9.11: This shows a two-line trailer that has been modified by the addition of a dual-relay emergency valve (G), a CA-type male coupling (E) and a dummy coupling (F) complete with retaining chain.

When towed by a two-line tractor the yellow service (control) line and the red emergency (supply) line are both connected to the trailer in the normal way, but the new CA-type coupling of the third line is connected to the dummy coupling (F).

When towed by a three-line tractor all three trailer couplings are used.

Figure 9.12: This shows a two-line trailer that has been modified by the addition of a relay valve (B), a change-over (non-return) valve (D), a CA-type male coupling (E) and a dummy coupling (F) complete with retaining chain.

When this trailer is drawn by a two-line tractor the yellow (service) control line and the red emergency (supply) line are connected in the normal way, but the new CA-type coupling is connected to the dummy coupling.

When this trailer is drawn by a three-line tractor, all three connections are used.

The Department of Transport have produced three documents on this subject:

1 *HGV Brake Systems: Braking Connections for Goods Vehicles and Trailers*, intended for fleet engineers, etc;
2 a leaflet *3 into 2 can GO*, advice for drivers;
3 a wall chart *3 into 2 can GO*, suitable for the workshop and transport area.

These are free and the address from which to obtain literature is shown on page 150.

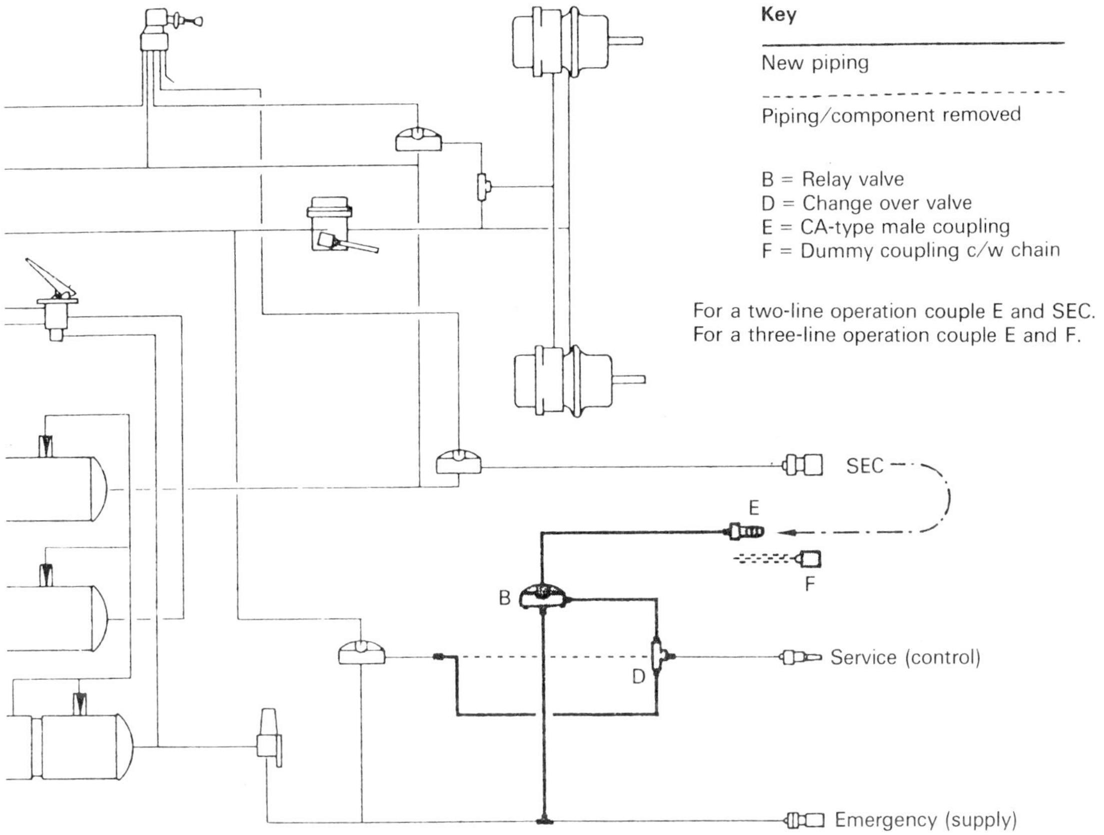

Figure 9.10 Tractor two/three-line conversion (2)

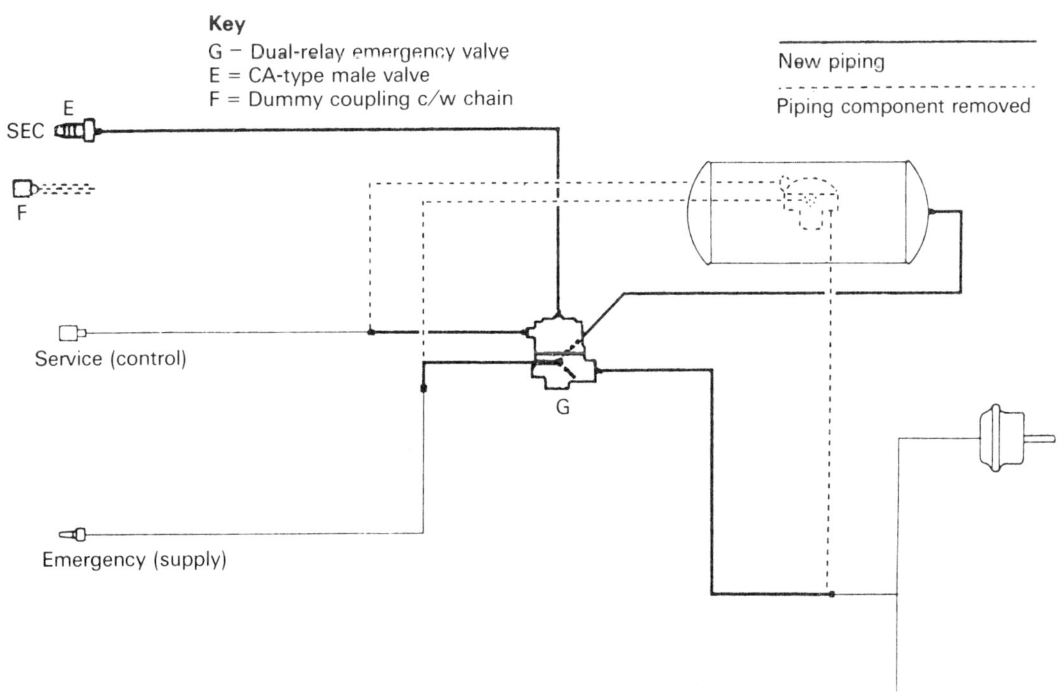

Figure 9.11 Trailer two/three-line conversion (1)

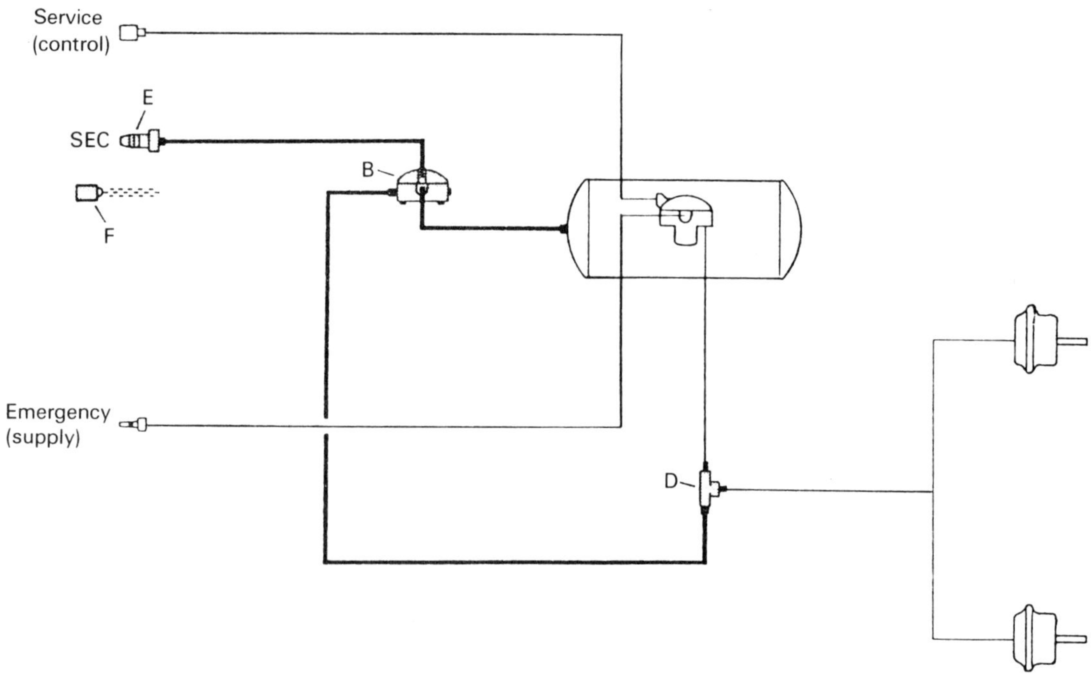

Figure 9.12 Trailer two/three-line conversion (2)

10 ISO, BSI and DIN symbols

Within the engineering industry world-wide, standardisation has been the aim for many years. Some success has been achieved, but much has still to be done. One needs only to cite the aim of a unified thread as being the one and only thread of the future. Many years have elapsed since this was the aim and thread types are just as varied as ever (if not even more varied than years ago). However, praise must be given for the success achieved by a number of international standards organisations. Here we will deal with three, although this does not depreciate the work of the other organisations.

In the UK, the British Standards Institution (BSI) is well-known and they act in this country on behalf of two other organisations, the International Organisation for Standardisation (ISO), which is based in Geneva, and Deutsches Institüt für Normung (DIN), based in West Germany. The aspects of their work that are of particular interest, relate to the numbering of ports on valves, to indicate where air flow is going to and coming from. In addition, they have devised a code of symbols to represent pneumatic valves.

ISO 6786:1980(E)

This standard lays down identification numbers to identify ports on units. It consists of one or two-digit numbers, as follows:

First digit:

0 suction connection
1 energy inlet
2 energy supply (not to atmosphere, see digit 3)
3 exhaust
4 pilot connection (entry into unit)
5 spare
6 spare
7 anti-freeze liquid connection
8 lubricating oil connection (for compressor)
9 cooling liquid connection (for compressor).

Second digit:

A second digit may be added, when several similar connections are available (for example with split systems), but the allocation of the second digit is left to the discretion of the manufacturer. However, the following codes are not discretionary:

71 anti-freeze inlet
72 anti-freeze outlet
81 lubricating oil inlet
82 lubricating oil outlet
91 cooling liquid inlet
92 cooling liquid outlet.

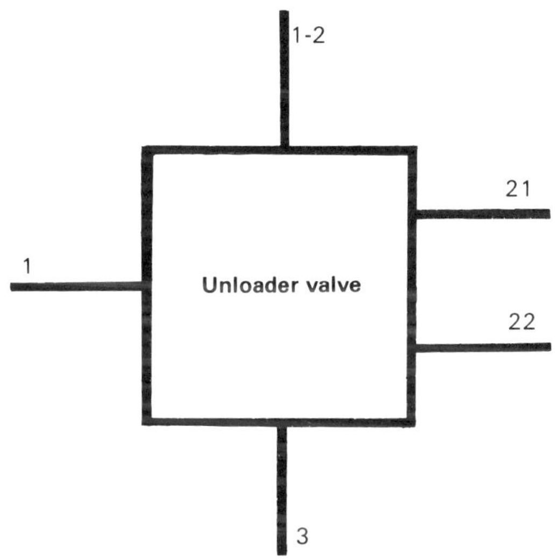

Source: BSI 2917:1977/ISO 1219: 1976)

Figure 10.1 Building block principle

Figure 10.1 shows, diagrammatically, the connections to an unloader valve:

1 inlet from compressor,
1-2 inlet, using a valve for filling the air brake system, and supply, by use of a valve, for inflating tyres,
3 vent to atmosphere,
21 supply to air reservoir,
22 supply (switching connection).

This is a specification for graphical symbols used on diagrams for fluid power systems and components. Although these are excellent, their value to those engaged in transport operations and maintenance is somewhat limited, due to the fact that they are common to hydraulic and pneumatic systems and basically intended for industrial equipment. Never-

Cylinder with cushion :		
– with single fixed cushion	Cylinder incorporating fixed cushion acting in one direction only	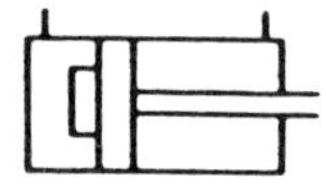
– with double fixed cushion	Cylinder with fixed cushion acting in both directions	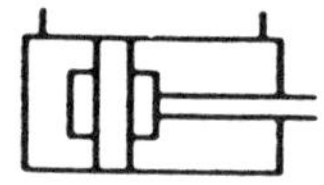
– with single adjustable cushion	The symbol is a combination of 6.5.4.1 and 5.2.3 (sloping arrow)	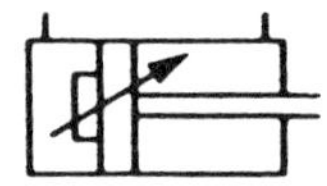
– with double adjustable cushion	The symbol is a combination of 6.5.4.2 and 5.2.3 (sloping arrow)	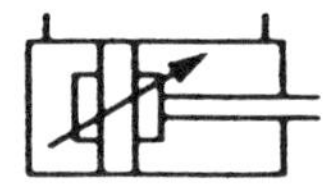
Directional control valve 4/2 : – controlled by pressure in both directions by means of a pilot valve (with a single solenoid and spring return)	Directional control valve with 4 ports and 2 distinct positions	Detailed 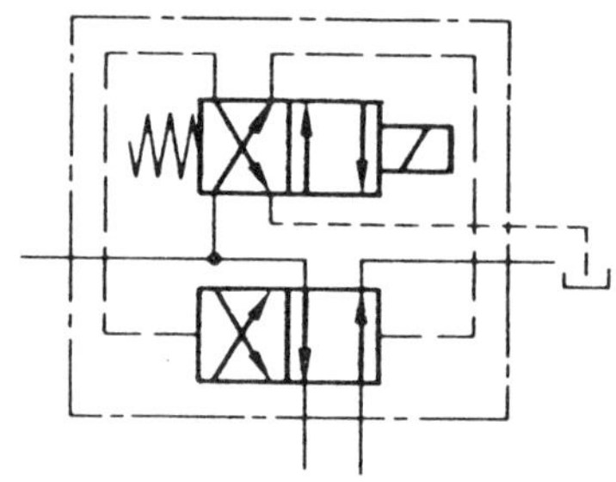 Simplified 
Directional control valve 5/2 : – controlled by pressure in both directions	Directional control valve with 5 ports and 2 distinct positions	

Figure 10.2 ISO/BSI symbols (1)

theless, the principles are sound, in that straight lines are used to represent pipelines, and curves and rectangles are used to represent basic valves. The complexity of diagrams becomes apparent when the rectangles are filled with the mechanism that is to be found in air brake equipment.

DIN 74253:1979

This document lists symbols, without the same internal details shown in the ISO/BSI document. It is a 25-page document, closely spaced, and it is therefore impractical to reproduce the entire contents. Thus, a small, random selection of such symbols is shown in Figure 10.4. These symbols are in general use in the instruction/service sheets and booklets that are produced by European manufacturers. Currently, most UK manufacturers are including the relevant symbol on the service sheets produced in relation to each of their valves.

Figure 10.5 shows a layout diagram (produced by Robert Bosch Ltd) representing an HGV with air suspension, using DIN symbols.

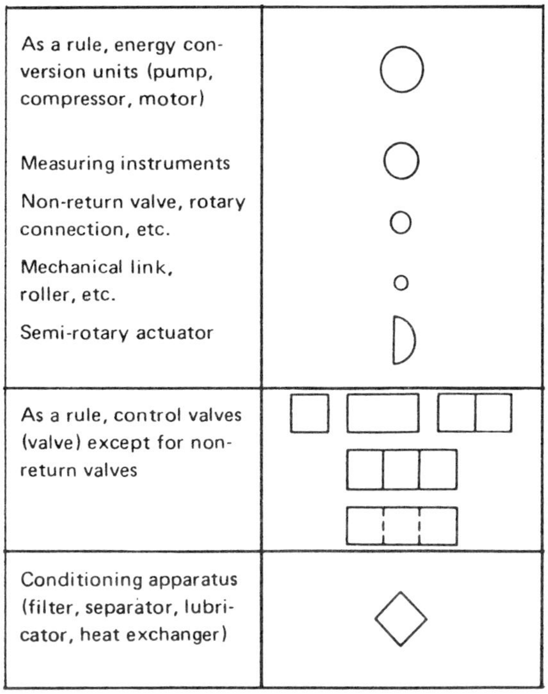

Figure 10.3 ISO/BSI symbols (2)

	Compressor	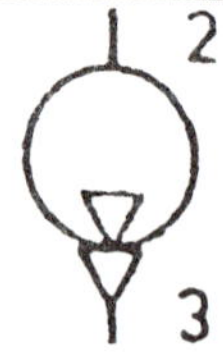	Vacuum Pump
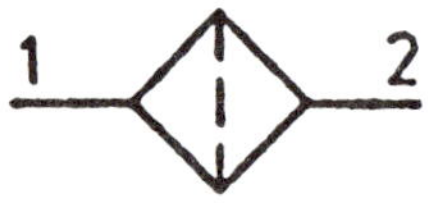	Line Filter		Anti Freezer–Pneumatic Trigger
	Drain Valve		Auto Drain Valve
	Rods & Mechanical Links		Test Point (In Line)
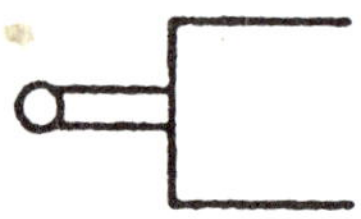	Operated by Rod	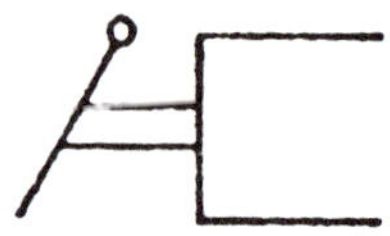	Operated by Hand Lever
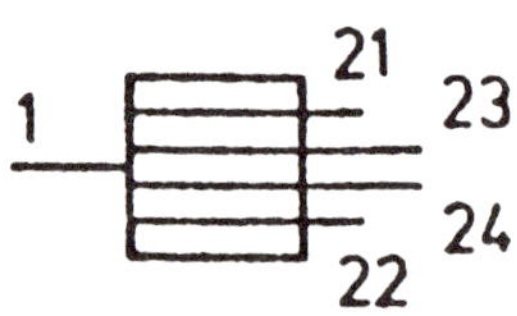	Four Circuit Protection Valve	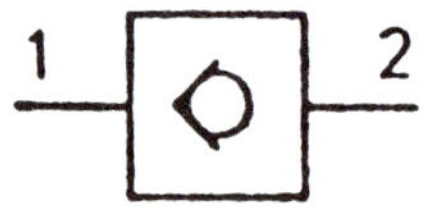	Non Return Valve
	Pressure Regulator	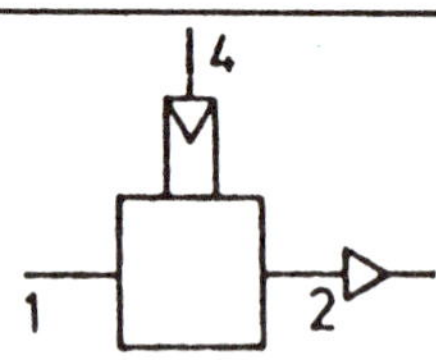	Relay Valve
	Pressure Gauge		Stop Cock
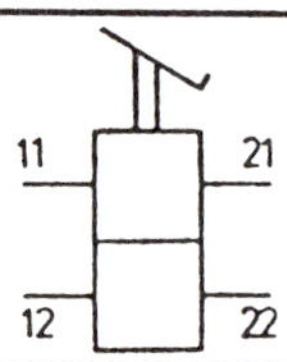	Foot Valve - 2 Circuit	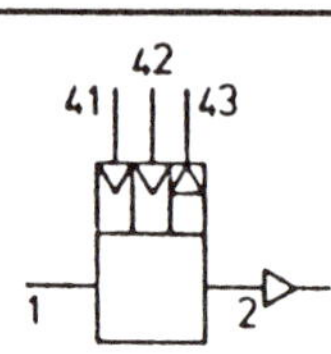	Trailer Control Valve 3 Input

Figure 10.4 DIN symbols

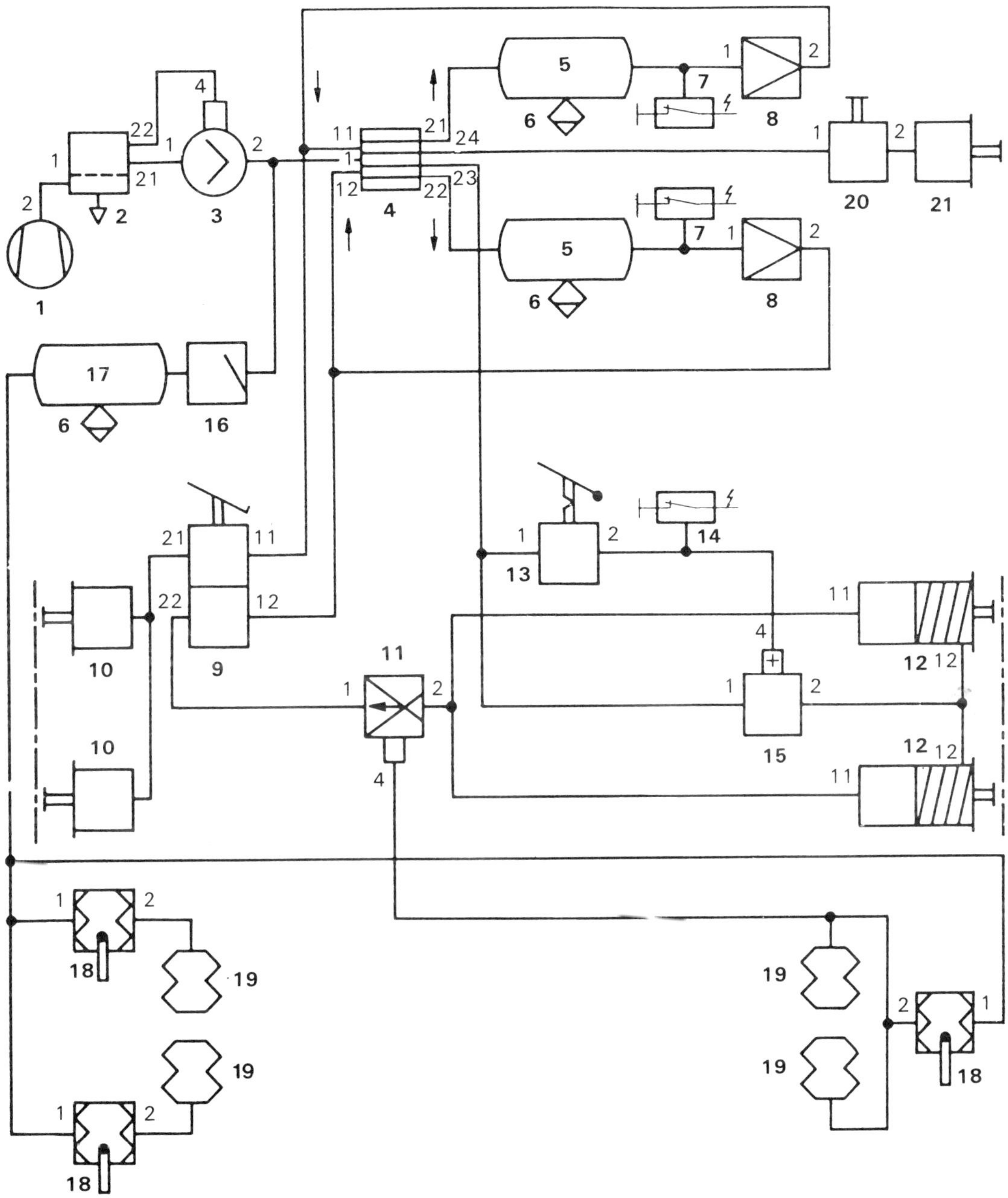

Note: Bold numbers on the diagram refer to the key. The lighter numbers refer to routing of air.

Key

1 = Air compressor
2 = Pressure regulator
3 = De-icing pump
4 = Four-circuit protection valve
5 = Air reservoir (brakes)
6 = Drain valve
7 = Low-pressure indicator
8 = Pressure limiter
9 = Service brake valve
10 = Brake cylinder
11 = Variable-load valve
12 = Spring brake
13 = Parking brake valve
14 = Stop-light switch
15 = Relay valve
16 = Pressure-protection valve
17 = Air reservoir (suspension)
18 = Height control valve
19 = Air suspension bellows
20 = Push-button valve (auxiliary load)
21 = Working cylinder (auxiliary load)

Figure 10.5 HGV with air suspension

Useful addresses

Bendix Ltd, Douglas Road, Kingswood, Bristol BS15 2NL, UK.
BHVSI, Bendix Heavy Vehicle Systems Italia (formerly Magnetti Marelli), Balaclava Road, Fishponds, Bristol BS16 3LJ, UK.
Robert Bosch Ltd, PO Box 98, North Orbital Road, Denham, Uxbridge UB9 5HJ, UK.
British Standards Institution, 2 Park Street, London W1A 2BS, UK.
Chartered Institute of Transport, 80 Portland Place, London W1, UK.
V L Churchill Ltd, PO Box 3, London Road, Daventry, Northants, UK.
Clayton Dewandre Co Ltd, Service Division, Mill Road, Rugby CV12 1BZ, UK.
Paul Dahl Ltd (*see* Garphyttan).
Department of Transport Literature available from: DOE/DTp Stores, Building 3, Victoria Road, South Ruislip, Middx HA4 0NZ, UK;
The Vehicle Inspectorate - Certification Agency, 1 The Eastgate Office Centre, Eastgate Road, Bristol, BS5 6XX, UK.
ERF Ltd, Service Centre, Middlewich, Cheshire CW10 0JZ, UK.
Feeny & Johnson Ltd, Alperton Lane, Wembley, Middx HA0 1JJ, UK.
Freight Transport Association, Hermes House, St John's Road, Tunbridge Wells, Kent TW4 9UZ, UK.
Garphyttan Ltd (Haldex), Hilton Road, Aycliffe Industrial Estate, Newton Aycliffe, Co Durham DL5 6SX, UK.
Grau Ltd, Glenfield Park, Philips Road, Blackburn BB1 5PF, UK.
Hope Technical Developments Ltd, High Street, Ascot, Berks, UK.
Institution of Mechanical Engineers, 1 Birdcage Walk, London, SW1, UK.
Institute of Road Transport Engineers, 1 Cromwell Place, London SW7 2JF, UK.
Iveco Ford Truck Ltd, Langley Assembly Plant, Sutton Lane, Langley, Slough SL3 8AQ, UK.
Knorr-Bremse GmbH (*see* Garphyttan Ltd (Haldex)).
Magnetti Marelli Ltd (*see* BHVSI).
Road Haulage Association, Roadway House, 104 New King's Road, Fulham, London SW6 4LN, UK.
Rubery Owen-Rockwell Ltd, Rackery Lane, Llay, Nr Wrexham, Clwyd LL12 0PB, UK.
SAB (*see* Garphyttan Ltd (Haldex)).
VL Bus and Coach Co (UK) Ltd, Heathcote, Warwick CV34 6LX, UK.
WABCO Ltd (*see* Clayton Dewandre Ltd).
York Trailer Co Ltd, Yafforth Road, Northallerton DL7 8UD, UK.

Index